سینا زندگی فارجی

طلاکوب روی جلد وقف زده شود

D

The McGraw-Hill Infrastructure Series

PLANNING

DICKEY *Metropolitan Transportation Planning, 2/e (1983)*
HORONJEFF & McKELVEY *Planning and Design of Airports, 3/e (1983)*
HUTCHINSON *Principles of Urban Transport Planning (1974)*
KANAFANI *Transportation Demand Analysis (1983)*
MORLOCK *Introduction to Transportation Engineering and Planning (1978)*

ANALYSIS & INSPECTION

BAKHT & JAEGER *Bridge Analysis Simplified (1985)*
BOWLES *Foundation Analysis and Design (1982)*
LAURSEN *Structural Analysis (1978)*
LEET *Reinforced Concrete Design (1984)*
NAAMAN *Prestressed Concrete Analysis and Design (1982)*
RAU & WOOTEN *Environmental Impact Analysis Handbook (1979)*
SACK *Structural Analysis (1984)*
WILLIAMS & LUCAS *Structural Analysis for Engineers (1967)*
WINTER & NILSON *Design of Concrete Structures, 9/e (1979)*

ENGINEERING & DESIGN

BILLINGTON *Thin-Shell Concrete Structures, 2/e (1982)*
BRUSH & ALMROTH *Buckling of Bars, Plates, and Shells (1975)*
FULLER *Engineering of Pile Installations (1983)*
GAYLORD & GAYLORD *Structural Engineering Handbook (1979)*
HALL & DRACUP *Water Resource Systems Engineering (1970)*
LePATNER & JOHNSON *Structural and Foundation Failures: A Casebook for
   Architects, Engineers, and Lawyers (1982)*
KREBS & WALKER *Highway Materials (1971)*
QUINN *Design and Construction of Ports and Marine Structures (1972)*
RINALDI *Modeling and Control of River Quality (1978)*
ROBINSON *Highways and Our Environment (1971)*
WALLACE & MARTIN *Asphalt Pavement Engineering (1967)*
YANG *Design of Functional Pavements (1973)*

BRIDGE
ANALYSIS
SIMPLIFIED

BAIDAR BAKHT

Principal Research Engineer
Ministry of Transportation and Communications, Ontario

LESLIE G. JAEGER

Professor of Civil Engineering and Applied Mathematics
The Technical University of Nova Scotia

McGRAW-HILL BOOK COMPANY

New York   St. Louis   San Francisco   Auckland
Bogotá   Hamburg   Johannesburg   London
Madrid   Mexico   Montreal   New Delhi
Panama   Paris   São Paulo   Singapore
Sydney   Tokyo   Toronto

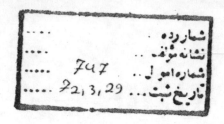

To Anita and Kathleen

Library of Congress Cataloging in Publication Data

Bakht, Baidar.
  Bridge analysis simplified.

  Includes index.
  1. Bridges—Design.   I. Jaeger, Leslie G.   II. Title.
TG300.B34   1985   624'.25   84-28919
ISBN 0-07-003020-0

234567890   DOC/DOC   8987

ISBN 0-07-003020-0

The editors for this book were Joan Zseleczky and Nancy Warren, the designer was Naomi Auerbach, and the production supervisor was Teresa F. Leaden. It was set in Caledonia by Progressive Typographers.

Printed and bound by R. R. Donnelley & Sons Company.

CONTENTS

v

About the Authors

BAIDAR BAKHT has had years of first-hand experience in load distribution and other aspects of bridge engineering as the manager of the structural research office for the Ministry of Transportation and Communications of Ontario. He played a key role in the development of the Ontario Highway Bridge Design Code and is a recipient of the Moisieff Award from the American Society of Civil Engineers. His active research in bridge analysis has led to frequent contributions to the technical literature on simplified bridge analysis. He is well known for his work in bridge testing and evaluation. He obtained his M.Sc. from London University.

LESLIE G. JAEGER is a professor of civil engineering and applied mathematics at the Technical University of Nova Scotia. The chairman of the Methods Analysis Committee, he is also technical editor of the Ontario Highway Bridge Design Code. A recipient of the A. B. Sanderson Award from the Canadian Society for Civil Engineering, and a Fellow of the Royal Society of Edinburgh and the Canadian Society for Civil Engineering, he is the author of numerous books and research publications on bridge design and analysis. He obtained his Ph.D. from London University.

PREFACE

Most highway bridges in North America are designed by the American Association of State Highway and Transportation Officials (AASHTO) specifications. A principal assumption underlying the analysis methods of AASHTO is that bridges of a given type (e.g., slab-on-steel girder bridges) all behave similarly in their live-load distribution properties.

By contrast with the North American tradition, bridge analysis in Europe tends to be highly analytical and, usually, computer-based. Such methods as the grillage analogy method, the orthotropic plate method, and the finite element and finite strip methods are extensively used.

This book presents a number of simple methods of analysis which expand upon the AASHTO approach, making it consistent with the more refined European methods. The simple methods presented in this book are derived from the results of computer-based rigorous analyses.

The book is intended to be useful both to the practicing bridge engineer and to the engineering student. It thus contains both "know-how" and "know-why" material. For the most part the know-why material is to be found in Chapter 1, while the remaining chapters constitute a know-how treatment which is virtually complete in itself.

The book is also intended to bring home to the reader the physical behavior of bridges of different types and to help the designer in establishing a "feel" for the mechanisms of load distributions.

A Note about Units

In order to make this book as widely useful as possible, all measurements in it are given in one of two ways. If the context is that of the AASHTO specifications, then the measurements are given in the units of the United States Customary System (USCS), followed by their metric equivalents in parentheses. If the context is that of the Ontario Highway Bridge Design Code, the measurements are given in metric units, followed by their USCS equivalents in parentheses. In figures and tables, however, only one of the two systems of units is followed.

Baidar Bakht

Leslie G. Jaeger

ACKNOWLEDGMENTS

Much of the research upon which this book is based was carried out for the Ministry of Transportation and Communications of Ontario, Canada, in the process of the development of the Ontario Highway Bridge Design Code. The Ministry's help in the preparation of many of the drawings, and the permission of the American Society of Civil Engineers and the *Canadian Journal of Civil Engineering* for the reproduction of others, are gratefully acknowledged. The research also benefited greatly from grants awarded to one of the authors by the Natural Sciences and Engineering Research Council of Canada. This support is also gratefully acknowledged.

We are indebted to a number of people for their help in the preparation of this book, but especially to Mrs. Pam Mulcahy, who typed the manuscript, and to Mr. Barry Mitchell, who prepared many of the drawings.

Baidar Bakht

Leslie G. Jaeger

1

PRINCIPLES OF LOAD DISTRIBUTION

1.1 Introduction

Within a time span of approximately 30 years, from roughly 1950 to 1980, the science of bridge analysis has undergone major change. Following the advent of the digital computer, and the consequent development of analytical techniques based upon its use, the bridge designer has available today a number of powerful analytical tools in the so-called refined methods of analysis, including the following:

1. The grillage analogy method
2. The orthotropic plate method
3. The articulated plate method
4. The finite element method, including its finite strip formulation

These refined methods are by now well established for the analysis of load distribution in bridges of various types. References 7 to 9, 11, 14, and 25 are representative of a large technical literature.

A principal objective of this book is to provide a simplified approach to the use of these tools so that their fruits may be made available to the designer without the need for performing complicated analysis in the design office. Thus, the term *simplified methods*, as used repeatedly in later chapters, means the representation of the results of complicated analysis in simple form; it does not mean the use of some overly simplifying assumption about bridge behavior, such as treating the bridge as a simple beam. The simplified methods have been deliberately put into a form so that the sequences of steps which the designer will follow have a marked resemblance to the familiar AASHTO (American Association of State Highway and Transportation Officials) methods which have been in use in North America for many years.

It is important to realize that most of the refined methods have limitations as to the kinds of bridge superstructures which they are capable of representing. For example, the usual grillage analogy and the orthotropic plate methods are suitable for the analysis of bridges in which load distribution takes place mainly through flexure and torsion in the longitudinal and transverse directions, with deflections due to shear being negligibly small. Bridge types which fall within this category include the "shallow superstructure" group, i.e., the solid slab, voided slab, and slab-on-girder types shown in Fig. 1.1. Significantly absent from this group is the cellular or multicell type shown in Fig. 1.2a. A bridge having a cross section of this type experiences significant deformation due to shear, which is accompanied by bending of the top and bottom flanges about their own centerlines in the manner shown in Fig. 1.2b. For this reason, if the grillage or the orthotropic plate representation is to be employed for the analysis of a multicellular type of bridge, the grillage or orthotropic plate must be different from the normal, and must include provision for significant deflection due to shear. The so-called shear-weak orthotropic plate, which is introduced later in this chapter, meets this need.

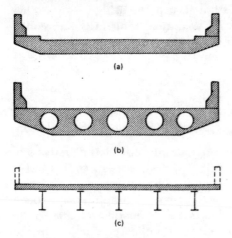

(a)

(b)

(c)

Figure 1.1 Cross sections of shallow superstructures: (*a*) a slab bridge; (*b*) a voided slab bridge; (*c*) a slab-on-girder bridge.

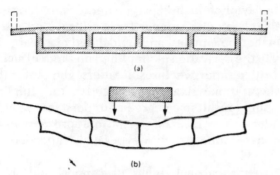

Figure 1.2   A cellular bridge: (*a*) cross section; (*b*) deflected cross section under concentrated loads.

In contrast to the grillage analogy and orthotropic plate methods the representation of a bridge as an articulated plate is appropriate when transverse distribution of load occurs mainly through shear forces, with little or no involvement of transverse bending stiffness. Figure 1.3*a* and *b* shows two bridge types which can be represented as articulated plates, whereas Fig. 1.3*c* shows the idealization involved, which is comprised of a number of longitudinal beams freely hinged together along their mating edges.

The finite element method, properly handled, is capable of representing bridge superstructures of all types. The particular kind of finite

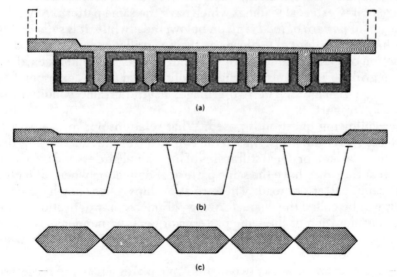

Figure 1.3   Cross sections of structures in which load distribution takes place mainly through transverse shear: (*a*) a multibeam bridge; (*b*) a multispine bridge; (*c*) idealized articulated plate.

element in a given case must, however, be chosen with care in order to represent properly the behavior of the bridge type under consideration.

In the remainder of this chapter, the properties of grillages, orthotropic plates, shear-weak orthotropic plates, and articulated plates are developed as a preliminary to the derivation of simplified methods of analysis for various types of bridge superstructures. Readers who prefer the mathematical kind of structural analysis are encouraged to study this first chapter carefully. This should result in a degree of understanding sufficient to enable the reader to use the same generic approach and to develop simplified methods for superstructure types which are not dealt with specifically in this book.

Readers for whom mathematical analysis has little appeal, and whose principal interests lie in the acquisition of design tools, may safely pass over the first chapter in a cursory way. The remaining chapters of the book give immediately applicable design methods which cover most cases.

1.2   The Concept of the Characterizing Parameter

Many simplified methods of bridge analysis are based on the concept of the *characterizing parameter*. This concept may conveniently be explained with the aid of Fig. 1.4.

Two grillages are shown in Fig. 1.4a and b. Each has the same number of longitudinal girders and the same number of transverse beams. The grillages carry external loadings which have the same pattern. A precise definition of *pattern of load* is given below; meanwhile, it is sufficient to note that the positions of the loads on the two grillages correspond so far as fraction of span and fraction of width are concerned, and that a given load on grillage 1 is a constant multiple of the load at the corresponding position on grillage 2. The two grillages have the same conditions of boundary support.

The following question is posed: What relationships must exist between the structural properties of the two grillages (i.e., such properties as the flexural and torsional stiffnesses of individual girders and beams) in order that they may have the same pattern of deflection when subjected to the same pattern of load? These relationships, once they have been found, will be called the characterizing parameters for deflection. Similarly, consideration of the same patterns of bending moments, twisting moments, etc., will lead to the characterizing parameters for these structural responses.

Figure 1.4c and d shows two orthotropic plates which are subjected to the same pattern of load. The same question is posed for these two as was posed above for grillages.

Before answering these questions, which are addressed in Secs. 1.4 and 1.5 below, it is appropriate to give some definitions.

Corresponding Plane Structures

Two plane structures are said to *correspond* if they have the same conditions of boundary support, and if the planform of one can be made to coincide with that of the other by application of (possibly different) scaling factors in two directions at right angles.

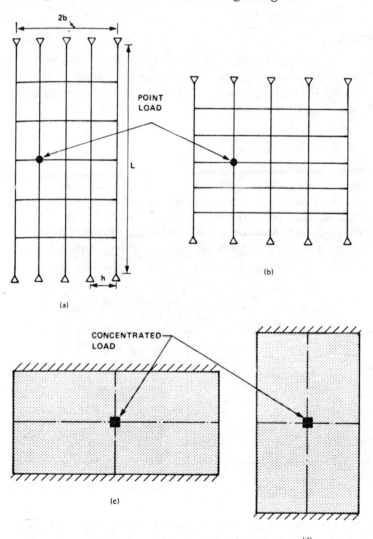

Figure 1.4 Plans for grillages and orthotropic plates: (*a*) grillage 1; (*b*) grillage 2; (*c*) orthotropic plate 1; (*d*) orthotropic plate 2.

For example, any two rectangles correspond in terms of the definition just given, even though their aspect ratios (length divided by breadth) are different. For rectangular planforms it is then convenient to define nondimensional span and nondimensional width coordinates in the manner shown in Fig. 1.5. In Fig. 1.5a an x coordinate runs from 0 to L and a y coordinate from 0 to $2b$, where L is span and b is half width. Then by defining $x' = x/L$ and $y' = y/b$, one obtains the nondimensional scheme of Fig. 1.5b, in which x' runs from 0 to 1 and y' runs from 0 to 2. This nondimensional scheme will be used consistently from now on.

Corresponding Points

Points in two plane structures correspond if they coincide when the planform of one is made to coincide with that of the other as defined above. Specifically, for rectangular planforms, the points correspond if they have the same nondimensional coordinates (x', y').

Pattern of Load

If in a structure number 1 two points, say a_1 and b_1, are identified, and if in a corresponding structure number 2 the corresponding two points a_2 and b_2 are identified, then the patterns of load are defined to be identical if the ratio of load intensity at a_1 to that at a_2 is the same as the ratio of load intensity at b_1 to that at b_2 for all pairs of corresponding points in the two structures.

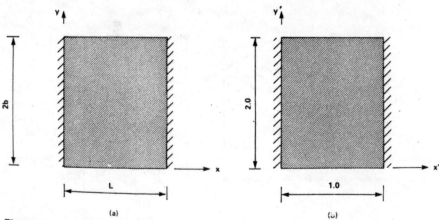

(a)   (b)

Figure 1.5 Coordinate systems for orthotropic plate planforms: (a) dimensional coordinate system; (b) nondimensional coordinate system.

Characterizing Parameters

If in two corresponding structures the necessary and sufficient condition for identical patterns of distribution of a given structural response, as a consequence of applying the same pattern of external load to the two, is that a certain one or more nondimensional parameters shall have the same value in the two structures, then those one or more parameters are defined to be the characterizing parameters for the structural response concerned.

1.3 The Use of Characterizing Parameters in Analysis

Provided that the number of characterizing parameters is not more than two, or in some cases three, the existence of characterizing parameters is of very great assistance in the development of simplified methods of analysis that are based upon results of often complex analyses. These characterizing parameters are used as axes of suitable design charts or tables. Having calculated the values of the characterizing parameters involved, the engineer can then identify a particular chart or table and read directly the value of the structural response concerned without performing any rigorous analysis. Clearly, a great economy of time and effort is achieved, if, for many structures such as highway bridges, much detailed analysis can be replaced by the calculation of the values of a few parameters, followed by the use of suitable charts.

Further, it is often the case that, for a particular kind of structure, the range of numerical values of its characterizing parameters becomes known to the designer to fall within certain well-recognized limits. The designer using such parameters in the analysis rapidly acquires a "feel" for the expected values of the parameter and thus for the kinds of distribution of the structural responses that are to be expected. With the help of the known ranges of a characterizing parameter, it is easy to ensure that a simplified method of analysis based on the parameter covers the whole range of the structural type concerned.

1.4 Characterizing Parameters for Grillages

A typical grillage, such as that shown in Fig. 1.4a, comprises a number of equal longitudinal girders, each having flexural stiffness EI, torsional stiffness GJ, and length L. The longitudinals are spaced a distance h apart and are interconnected by a number of equally spaced transverse beams each of which has flexural stiffness EI_T and torsional stiffness GJ_T.

If such a grillage is subjected to a given external loading, the problem of finding the responses may be expressed in the form

$$[K]\{a\} = \{W\} \tag{1.1}$$

in which $\{W\}$ is a vector of externally applied forces, $\{a\}$ is a vector whose elements depend upon the deflections and rotations of the nodes and have the dimensions of force, and $[K]$ is a matrix whose elements have no dimensions.

It is readily verified that the elements of $[K]$ depend upon the following nondimensional parameters:

$$\left[\frac{L}{h}\right]^3\left[\frac{EI_T}{EI}\right] \quad \left[\frac{L}{h}\right]\left[\frac{GJ_T}{EI}\right] \quad \text{and} \quad \left[\frac{h}{L}\right]\left[\frac{GJ}{EI_T}\right]$$

If two different grillages have the same pattern of external load, then the vector $\{W\}$ is the same for both, within a simple scalar multiplier. If, further, the two grillages have the same values of the three nondimensional parameters just identified, then the $[K]$ matrix is the same for both. Hence the vector $\{a\}$ must be the same for both, within a simple scalar multiplier.

Once the vector $\{a\}$ is known, the patterns of distribution of deflections, bending moments, twisting moments, and shear forces follow directly. Hence, it is concluded that the three nondimensional parameters given above are the characterizing parameters for all these structural responses.

For purposes of later comparison with orthotropic plate behavior, it is convenient at this point to express the grillage parameters in terms of an equivalent orthotropic plate, which is obtained by distributing the stiffnesses of the individual girders uniformly across the width and the stiffnesses of the transverse beams uniformly along the length. This means, for example, that the flexural stiffness EI of a girder is written as

$$EI = D_x h \tag{1.2}$$

where D_x is longitudinal flexural stiffness per unit width of the equivalent orthotropic plate. Three suitable nondimensional parameters for the grillage, expressed in equivalent orthotropic plate form, are then

$$\left(\frac{L}{h}\right)^4\left(\frac{D_y}{D_x}\right) \quad \left(\frac{L}{h}\right)^2\left(\frac{D_{yx}}{D_x}\right) \quad \text{and} \quad \left(\frac{h}{L}\right)^2\left(\frac{D_{xy}}{D_y}\right)$$

where plate rigidities are as defined in Sec. 2.2. The quantity h/L can be eliminated from the second and third of these by multiplying or dividing, as the case may be, by the square root of the first one. Further, inverting the first parameter and then taking the fourth root gives a parameter in which the ratio h/L has the first power. The result of these adaptations is to define the following:

$$\alpha_1 = \frac{D_{xy}}{2(D_x D_y)^{0.5}} \tag{1.3}$$

$$\alpha_2 = \frac{D_{yx}}{2(D_x D_y)^{0.5}} \tag{1.4}$$

$$\theta = \frac{h}{L}\left(\frac{D_x}{D_y}\right)^{0.25} \tag{1.5}$$

If two different grillages have the same values of α_1, α_2, and θ, then the two grillages will have the same patterns of distribution of deflections, shearing forces, and bending and twisting moments when subjected to the same pattern of load. These three parameters are the characterizing parameters for grillage behavior.

1.5 Characterizing Parameters for Orthotropic Plates

It is shown in books on the theory of plates, for example, in Ref. 24, that the deflection w of an orthotropic plate is governed by the following equation:

$$D_x \frac{\partial^4 w}{\partial x^4} + (D_{xy} + D_{yx} + D_1 + D_2) \frac{\partial^4 w}{\partial x^2 \partial y^2} + D_y \frac{\partial^4 w}{\partial y^4} = q(x, y) \tag{1.6}$$

Since it is *patterns* of deflection that are being sought, Eq. (1.6) is recast in terms of dimensionless quantities $x' = x/L$ and $y' = y/b$, where L is the span and b is the half width, as shown in Fig. 1.5. Then

$$\frac{\partial}{\partial x} \equiv \frac{1}{L}\frac{\partial}{\partial x'} \tag{1.7}$$

and

$$\frac{\partial}{\partial y} \equiv \frac{1}{b}\frac{\partial}{\partial y'} \tag{1.8}$$

so that Eq. (1.6) gives

$$\frac{D_x}{L^4}\frac{\partial^4 w}{\partial x'^4} + \left(\frac{D_{xy} + D_{yx} + D_1 + D_2}{L^2 b^2}\right)\frac{\partial^4 w}{\partial x'^2 \partial y'^2} + \frac{D_y}{b^4}\frac{\partial^4 w}{\partial y'^4} = \phi(x', y') \tag{1.9}$$

where $\phi(x', y')$ is the expression of the externally applied load in terms of x' and y'.

Multiplying through by $L^2 b^2/(D_x D_y)^{0.5}$ yields

$$\theta^2 \frac{\partial^4 w}{\partial x'^4} + 2\alpha \frac{\partial^4 w}{\partial x'^2 \partial y'^2} + \frac{1}{\theta^2} \frac{\partial^4 w}{\partial y'^4} = \frac{L^2 b^2}{(D_x D_y)^{0.5}} \phi(x', y') \qquad (1.10)$$

where

$$\alpha = \frac{D_{xy} + D_{yx} + D_1 + D_2}{2(D_x D_y)^{0.5}} \qquad (1.11)$$

$$\theta = \frac{b}{L}\left(\frac{D_x}{D_y}\right)^{0.25} \qquad (1.12)$$

Two different orthotropic plates will have the same right-hand side for Eq. (1.10), to within a simple scalar multiplier, provided that they have the same *pattern* of externally applied load as defined earlier. The two plates will have the same left-hand side of Eq. (1.10), and therefore the same patterns of deflection, provided that they have the same values of α and θ as defined by Eqs. (1.11) and (1.12).

It is concluded that α and θ are the characterizing parameters for the deflections of rectangular orthotropic plates.

Turning to longitudinal bending moments per unit width, Ref. 24 gives

$$M_x = -\left(D_x \frac{\partial^2 w}{\partial x^2} + D_1 \frac{\partial^2 w}{\partial y^2}\right) \qquad (1.13)$$

This equation may be expressed in terms of the dimensionless coordinates x' and y' as

$$M_x = -\frac{D_x}{L^2}\left[\frac{\partial^2 w}{\partial x'^2} + \frac{L^2}{b^2}\left(\frac{D_1}{D_x}\right)\frac{\partial^2 w}{\partial y'^2}\right] \qquad (1.14)$$

Since w is already known to be a function of α and θ, it is clear from Eq. (1.14) that M_x cannot be exactly characterized by only these two, there being present another nondimensional parameter, namely, $(L^2/b^2)(D_1/D_x)$.

However, provided that the effect of the coupling rigidity D_1 is small, it remains possible that the parameters α and θ will characterize M_x sufficiently closely for design purposes in bridges. Extensive analysis of a large number of cases has confirmed that this is indeed the case.

It is very instructive at this stage to compare the characterizing parameters for grillages, as given in Eqs. (1.3) to (1.5), with those for orthotropic plates as given in Eqs. (1.11) and (1.12). One notes immediately that Eqs. (1.5) and (1.12) are equivalent since, for a grillage with a given number of longitudinals, the half width b is related by a simple numerical multiplier to the spacing h between girders. Further, if Eqs.

(1.3) and (1.4) are added together, there results

$$\alpha_1 + \alpha_2 = \frac{D_{xy} + D_{yx}}{2 \, (D_x D_y)^{0.5}} \tag{1.15}$$

Equation (1.15) is the same as Eq. (1.11), except that the coupling rigidities D_1 and D_2 are absent from the right-hand side. This is an expected outcome since the coupling rigidities arise from the Poisson's ratio effects associated with the two-way bending of the plate, effects which are absent from grillage behavior. Indeed, if one replaces a grillage by its orthotropic plate analogue, in which the flexural and torsional rigidities of the individual grillage members are uniformly distributed, then the α for that equivalent orthotropic plate is precisely the right-hand side of Eq. (1.15).

The fact that in orthotropic plates not only deflections but also other effects such as longitudinal bending moments can be represented, sufficiently closely for design purposes, by only two parameters α and θ, leads one to hope that the same may be true of grillages. This will be so if the grillage behavior, which is known to depend on the three parameters given by Eqs. (1.3) to (1.5), turns out to be mainly dependent on $\alpha_1 + \alpha_2$ and θ, with little dependence on the values of α_1 and α_2 separately. In order to investigate this postulate, it is convenient to use as a vehicle a particular type of grillage.

The Semicontinuum Plate or Grillage

Figure 1.6 shows a hybrid concept in which longitudinal bending and torsional rigidities are concentrated in a number of discrete longitudinal members, while transverse bending and torsional rigidities are uniformly distributed along the length of the structure. This concept has been used extensively and is described in Ref. 12. It is a convenient concept for the

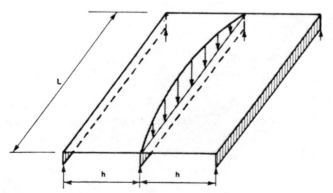

Figure 1.6   A simple semicontinuum grillage.

purposes of analysis, since it enables the analyst to express applied loads as continuous functions of the x coordinate. In particular, it lends itself to harmonic analysis of load effects. The three-girder semicontinuum shown in Fig. 1.6 was loaded along its center girder by a line load of the form $P \sin(\pi x/L)$ as shown. A three-girder grillage was chosen for this investigation because grillages with large numbers of longitudinals can be expected to approximate plate behavior more closely than those with small numbers; hence, if one is investigating possible differences between grillage behavior and orthotropic plate behavior, it is prudent to have a small number of longitudinals.

With the line loading shown, the total longitudinal bending moment at midspan, which is shared between the three longitudinals, is statically determinate and is given by

$$M_0 = \frac{PL^2}{\pi^2} \tag{1.16}$$

The bending moment accepted by the center girder at midspan may be expressed as ρM_0, where ρ is a distribution coefficient for longitudinal moments whose value depends upon α_1, α_2, and θ as given by Eqs. (1.3), (1.4), and (1.5), respectively.

The behavior of the distribution coefficient ρ is shown in Fig. 1.7. Three values of θ are taken: 0.0, 0.5, and 1.0. For each value of θ various values of α (= $\alpha_1 + \alpha_2$) are taken between 0.0 and 1.0. For each value, α is composed in three different ways, i.e., all α_1, half α_1 and half α_2, and all α_2. It will be seen that for a given value of θ the value of ρ responds mainly to the value of α and is only slightly influenced by the way in which α is built up between α_1 and α_2.

A significant conclusion, and a useful unifying one between grillages and orthotropic plates, may be drawn from the above: For both grillages and orthotropic plates, as applied to highway bridges, deflections and longitudinal moments are characterized, sufficiently accurately for design purposes, by

$$\alpha = \frac{D_{xy} + D_{yx} + D_1 + D_2}{2(D_x D_y)^{0.5}} \tag{1.17}$$

$$\theta = \frac{b}{L}\left(\frac{D_x}{D_y}\right)^{0.25} \tag{1.18}$$

where, in Eq. (1.17), $D_1 = D_2 = 0$ in the case of a grillage. The parameters α and θ have been used as the basis of a number of simplified methods; in particular, Ref. 2 gives the development of such a method for the Ontario highway bridge design code [17].

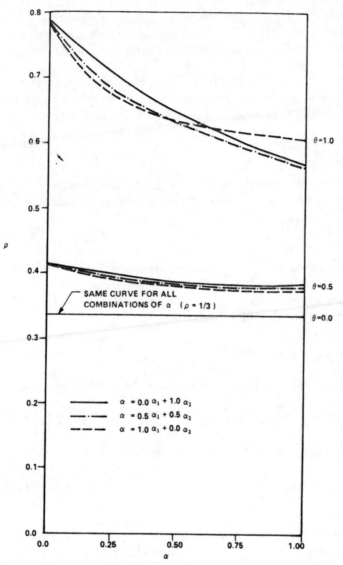

Figure 1.7   Effect of the ratio of α_1 and α_2 on ρ.

1.6   Characterizing Parameters for Articulated Plates

It was noted earlier with reference to Fig. 1.3 that there are certain kinds of bridges in which transverse bending stiffness is very small and load distribution takes place mainly by shear forces. This behavior leads to the consideration of a particular class of orthotropic plate in which the transverse flexural rigidity D_y and the transverse torsional rigidity D_{yx} both approach zero. The rigidities D_1 and D_2 also become zero in this case, since the two-way bending coupling action is eliminated. The result is an articulated plate.

As D_y approaches zero it is clear from Eq. (1.17) and (1.18) that both α and θ approach infinity. However, eliminating D_y between these two equations leads to the definition of a parameter, $\theta/\sqrt{\alpha}$, given by

$$\frac{\theta}{\sqrt{\alpha}} = \sqrt{2}\left(\frac{b}{L}\right)\left(\frac{D_x}{D_{xy}}\right)^{0.5} \tag{1.19}$$

The parameter $\theta/\sqrt{\alpha}$ as defined above is identical, except for a simple scalar multiplier, with the parameter β used in Ref. 23 and defined by

$$\beta = \pi\left(\frac{2b}{L}\right)\left(\frac{D_x}{D_{xy}}\right)^{0.5} \tag{1.20}$$

So far as the deflections of such a plate are concerned, it is not difficult to show that β (or equivalently $\theta/\sqrt{\alpha}$) is the characterizing parameter. With D_y, D_{yx} becoming indefinitely small, the governing differential equation for an orthotropic plate [Eq. (1.6)] reduces to

$$D_x \frac{\partial^4 w}{\partial x^4} + D_{xy} \frac{\partial^4 w}{\partial x^2 \partial y^2} = q(x, y) \tag{1.21}$$

In terms of dimensionless coordinates x' and y', this becomes

$$\frac{D_x}{L^4} \frac{\partial^4 w}{\partial x'^4} + \frac{D_{xy}}{L^2 b^2} \frac{\partial^4 w}{\partial x'^2 \partial y'^2} = \phi(x', y') \tag{1.22}$$

which, on multiplying by $L^2 b^2 / D_{xy}$, may be written

$$\frac{\beta^2}{4\pi^2} \frac{\partial^4 w}{\partial x'^4} + \frac{\partial^4 w}{\partial x'^2 \partial y'^2} = \frac{L^2 b^2}{D_{xy}} \phi(x', y') \tag{1.23}$$

For two different plates, the right-hand side of Eq. (1.23) will be the same for both, within a simple scalar multiplier, provided that the patterns of load are the same. The left-hand side of Eq. (1.23) will be the same for both provided that the two plates have the same value of β, which is thus the characterizing parameter for deflection.

In articulated plates, the single parameter β also characterizes longitudinal bending moment, longitudinal shear, and transverse shear, as may be illustrated in the following simple example. Figure 1.8a shows three girders each of length L and width h, and of flexural rigidity EI and torsional rigidity GJ. The girders are simply supported at their ends and are freely hinged along their abutting edges. The middle girder carries a line load P sin $(\pi x/L)$, and the end conditions are such that rotations of the girders around their longitudinal centerlines are prevented at $x = 0$ and $x = L$.

As shown in Fig. 1.8b, let the interactive line load between girders 1 and 2 be $p \sin (\pi x/L)$. By symmetry this is also the line load between girders 2 and 3. Further, let the deflections of girders 1 and 2 be given by

$$w_1 = a_1 \sin \frac{\pi x}{L} \tag{1.24}$$

$$w_2 = a_2 \sin \frac{\pi x}{L} \tag{1.25}$$

and let the rotation of cross sections of girder 1 around the longitudinal axis be given by

$$\phi_1 = c_1 \sin \frac{\pi x}{L} \tag{1.26}$$

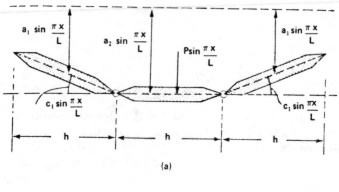

(a)

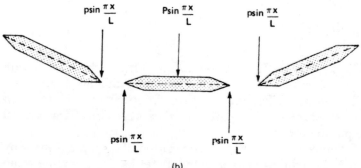

(b)

Figure 1.8 A three-girder articulated plate under symmetric first harmonic load. (a) Cross section of the structure. (b) Interactive line loading.

Then, from bending deflections of girder 1,

$$EI\frac{d^4w_1}{dx^4} = p \sin\frac{\pi x}{L} \tag{1.27}$$

whence, using Eq. (1.24),

$$ka_1 = p \tag{1.28}$$

where $k = EI\dfrac{\pi^4}{L^4}$

Similarly, from bending deflections of girder 2,

$$ka_2 = P - 2p \tag{1.29}$$

For the twisting of girder 1, consideration of an element of length δx gives

$$\frac{dT}{dx} = -\frac{h}{2}p \sin\frac{\pi x}{L}$$

where T = twisting moment = $GJ(d\phi_1/dx)$. Hence, using Eq. (1.26),

$$\frac{2\pi^2 GJ}{hL^2}c_1 = p \tag{1.30}$$

Deflection compatibility requires that

$$a_2 = a_1 + \frac{h}{2}c_1 \tag{1.31}$$

Equations (1.28) to (1.31) are sufficient for the determination of the unknowns p, a_1, a_2, and c_1 in terms of known quantities.
    Solving these, one finds

$$\frac{p}{P} = \frac{36}{108 + \beta^2} \tag{1.32}$$

$$\frac{P - 2p}{P} = \frac{36 + \beta^2}{108 + \beta^2} \tag{1.33}$$

The fractions given by Eqs. (1.32) and (1.33) are the distribution coefficients for deflection and longitudinal bending moment of girders 1 and 2 under first harmonic loading, and these fractions are functions of β only. Similarly, transverse and longitudinal shears are also characterized by the single parameter β.
    Figure 1.9 demonstrates the characterization of transverse shear by the parameter β. Two different articulated plates having the same value

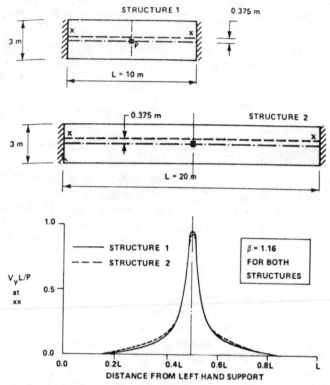

Figure 1.9 Characterization of transverse shear in articulated plates.

of β are seen to have the same distribution of $V_y L/P$ along a longitudinal cross section xx [5], where V_y is the intensity of transverse shear and P is the intensity of load.

It is worthy of note that a parameter equivalent to β (or $\theta/\sqrt{\alpha}$) is used in Ref. 22, where it is known as the stiffness parameter.

1.7 Characterizing Parameters for Longitudinal Bending Moments in Highway Bridges

Shallow Superstructure Types

The shallow superstructure types, as shown in Fig. 1.1, include the solid slab, voided slab, and slab-on-girder bridges. Extensive analysis has shown that the distribution of longitudinal moments in such bridges is given, sufficiently accurately for design purposes, by the grillage analogy or the orthotropic plate method.

The longitudinal bending moments are accordingly characterized by parameters α and θ as given by Eqs. (1.17) and (1.18). Figure 1.10 shows a typical (α, θ) chart from which may be read the necessary information relating to design for longitudinal bending moments. In this case the information given is the magnitude of D, a quantity which thereafter is used in a manner similar to that of the AASHTO specifications [1]. The quantity D is defined in Sec. 1.10.

Multicell Box Girders

In multicell box girder bridges, and also in voided slab bridges with large voids, transverse distribution of longitudinal bending moments is affected by deflections due to shear as well as deflections due to bending. The parameters α and θ, although still applicable, are then not capable of characterizing fully the distribution. It has been shown in Ref. 16 that an additional parameter, δ, is needed. This parameter is defined by

$$\delta = \frac{\pi^2 b}{L^2} \left(\frac{D_x}{S_y} \right)^{0.5} \tag{1.34}$$

where S_y is the transverse shear rigidity, being the product of the shear modulus and the equivalent shear area per unit length.

At first sight the fact that there are three parameters (α, θ, δ) needed for the characterization would lead one to surmise that the construction of design charts would be difficult. Fortunately, in the multicellular and voided slab bridges, the value of the parameter α varies hardly at all, being always near to 1.0. Hence the distribution of longitudinal bending moments is characterized, sufficiently accurately for design purposes, by the two parameters θ and δ.

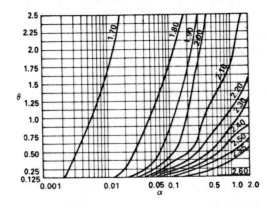

Figure 1.10 D values for three-lane bridges. (*From Ontario Highway Bridge Design Code, 1983.*)

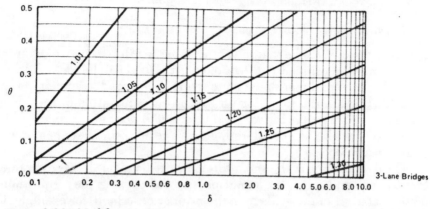

Figure 1.11 Modification factors F. (*From Ontario Highway Bridge Design Code, 1983.*)

Figure 1.11 shows a (θ, δ) chart from which a modification factor F is read. The modification factor is then applied as divisor to a value of D obtained as for a shallow superstructure type, so as to give a modified D which takes account of shear deformations.

Shear-Connected Beam Bridges

In shear-connected beam bridges, and also in laminated-wood bridges with nailed connections, the single parameter β characterizes longitudinal bending moments. One therefore has the option of preparing suitable design charts based upon this parameter. However, this is not necessary.

It is recalled that β is in fact $\theta/\sqrt{\alpha}$ to within a simple numerical multiplier, and that characterization by the parameter β corresponds to an orthotropic plate behavior in which D_y is small. In an (α, θ) chart such as the one shown in Fig. 1.10, various combinations of α and θ for this kind of orthotropic plate are possible.

For large values of α the contours of D in Figure 1.10 take the form $\theta/\sqrt{\alpha} = $ constant, which is itself a confirmation of the fact that $\theta/\sqrt{\alpha}$ (or, equivalently, β) is characterizing the behavior. Hence, instead of preparing a separate chart based upon the parameter β, one may use the existing (α, θ) chart, taking any convenient large value of α and reading the D value in the usual way. It is recommended that a value $\alpha = 2.0$ be used for this purpose. In this case, by use of Eqs. (1.19) and (1.20), the procedure is to calculate β from Eq. (1.20) and then obtain θ from $\theta = \beta/\pi$.

1.8   Characterizing Parameters for Longitudinal and Transverse Shears in Highway Bridges

Shallow Superstructure Types

It has been shown above that the patterns of shear forces in two different grillages carrying similar patterns of external load will be identical provided that the nondimensional parameters α_1, α_2, and θ defined in Eqs. (1.3) to (1.5) are the same for both. However, the attempt to reduce these three parameters to two (that is, α and θ) does not succeed when shear forces are being considered, unless a relationship between α_1 and α_2 is also available.

The dependence of the shear forces in a grillage on all three parameters α_1, α_2, and θ may be confirmed by considering the "equivalent plate" behavior of the grillage in the manner described, for example, in Ref. 24, the notation of which is widely adopted:

$$Q_x = \frac{\partial M_x}{\partial x} + \frac{\partial M_{yx}}{\partial y} \tag{1.35}$$

in which, for a grillage,

$$M_x = -D_x \frac{\partial^2 w}{\partial x^2} \tag{1.36}$$

and

$$M_{yx} = -D_{yx} \frac{\partial^2 w}{\partial x\, \partial y} \tag{1.37}$$

Equation (1.35) gives, using (1.36) and (1.37),

$$Q_x = -\left(D_x \frac{\partial^3 w}{\partial x^3} + D_{yx} \frac{\partial^3 w}{\partial x\, \partial y^2} \right) \tag{1.38}$$

which may be expressed in terms of the nondimensional coordinates x' and y' as

$$Q_x = -\frac{D_{yx}}{Lb^2}\left[\left(\frac{b}{L} \right)^2 \left(\frac{D_x}{D_{yx}} \right) \frac{\partial^3 w}{\partial x'^3} + \frac{\partial^3 w}{\partial x'\, \partial y'^2} \right] \tag{1.39}$$

The pattern of distribution of shear forces Q_x across a transverse cross section depends upon the expression in square brackets in Eq. (1.39). Thus, if two different grillages are to have the same pattern of distribution, they must have the same value of $(b/L)^2 (D_x/D_{yx})$. This means that they must have the same value of θ^2/α_2 as defined earlier. Since equality of the parameter θ for the two grillages has already been established as necessary, this means that α_2 must be the same for both.

Similarly, for the patterns of the shear forces Q_y to be the same, it is readily shown that α_1 must be the same for both.

If the shallow superstructure type of bridge is represented as an orthotropic plate instead of as a grillage, the treatment is essentially identical with that given above for grillages, and again leads to the conclusion that α_1, α_2, and θ are the characterizing parameters, that the pattern of the Q_x shear forces depends directly on α_2, that the pattern of the Q_y shear forces depends directly on α_1, and hence that reduction of the number of characterizing parameters from three to two does not appear possible.

An exception to the generality of the foregoing is made, however, when a known relationship exists between α_1 and α_2. For example, in a solid slab bridge, one has, very closely, $\alpha_1 = \alpha_2 = \alpha/2 = 0.5$. Hence, in this case the characterization is by θ only (or, equivalently, by β only).

Multicell Box Girders

The simplification given immediately above for solid slab bridges is available also in the case of multicell box girders for which, very closely, $\alpha_1 = \alpha_2 = \alpha/2 = 0.5$.

Shears in such bridges are thus characterized by θ and δ sufficiently closely for design purposes.

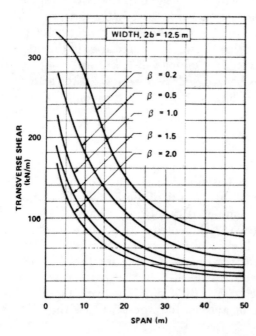

Figure 1.12 Transverse shear. (*From Ontario Highway Bridge Design Code, 1983.*)

Shear-Connected Beam Bridges

In shear-connected beam bridges the shears are characterized by the parameter β. Figure 1.12 shows how transverse shear force intensity due to the Ontario design loads may be read directly with respect to β and the span of a bridge.

1.9   Characterizing Parameters for Special Cases

Edge-Stiffened Plates

It has been shown in Ref. 3 that in torsionally soft orthotropic plates, the effect of two symmetrical edge beams on the distribution of longitudinal moments can be characterized by a parameter λ which is defined as follows:

$$\lambda = \frac{EI}{L}\left(\frac{1}{D_x^3 D_y}\right)^{0.25} \tag{1.40}$$

where EI is the flexural rigidity of each edge beam. It should be noted that the edge beam characterization is in addition to the (α, θ) characterization for unstiffened orthotropic plates. The implication of the λ characterization is that if symmetrical edge beams are attached to two bridges having the same values of α and θ, in such a way that the values of λ for the two bridges are also the same, then the pattern of distribution of longitudinal moments under similar load will be the same in the two bridges.

The λ idealization is more closely valid for torsionally soft bridges, e.g., slab-on-girder bridges, than for torsionally stiffer bridges. In the idealization analysis, it is assumed that the torsional rigidities of the edge beams themselves have negligible effect on load distribution.

The following expression has been developed in Ref. 20 for articulated plates, when the applied loading is analyzed into harmonics.

$$\lambda = \frac{2n\pi EI}{L}\left[\frac{1}{D_x(D_{xy} + D_{yx} + D_1)}\right]^{0.5} \tag{1.41}$$

where n is the number of the harmonic. Thus, the value of λ will change with the harmonic number under consideration.

It is important to recognize that the validity of the λ characterization is not limited to any harmonic number. If two bridges have the same value of λ for the first harmonic ($n = 1$), then clearly the two bridges will still have identical values of λ for higher harmonics. The characterization can, however, be conveniently based on the λ value for the first harmonic. It is of course recognized that the distribution pattern for moments due to different harmonics will be different. However, the distribution pattern for two bridges for a given harmonic will be the same

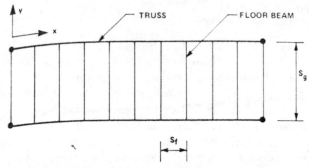

Figure 1.13 Plan of the floor system of a truss bridge.

provided that they both have the same value of λ as given by Eq. (1.41), using $n = 1.0$.

Floor Systems of Truss Bridges

For developing a simplified method to account for load distribution between the floor beams of a truss bridge floor system, the authors have formulated a single characterizing parameter ω which is given by

$$\omega = \left(\frac{S_g}{S_f}\right)\left(\frac{D_x}{D_y}\right)^{0.25} \tag{1.42}$$

where, as shown in Fig. 1.13, S_g is the center-to-center spacing between the two trusses, S_f is the spacing of floor beams, and D_x and D_y have their usual meanings as the uniformly distributed flexural rigidities in the longitudinal and transverse directions of the bridge, respectively.

The parameter ω is based on the premise that the torsional rigidities of the floor beams and of the load-transferring medium between the floor beams have negligible effect on the load distribution. It is noted that ω is also applicable to floor beams of two-girder bridges, in which case S_g is the center-to-center spacing of girders. Because of the formulation of a single characterizing parameter, the problem of load distribution analysis in floor beams of truss bridge floor systems is reduced to a single chart, as shown in Sec. 12.4.

1.10   Review of Existing Simplified Methods

Some of the commonly used simplified methods of bridge analysis are discussed in this section.

Distribution Coefficient Methods
Based upon Orthotropic Plate Theory

One of the earlier distribution coefficient methods is due to Guyon [10] and Massonnet [15]. This method is based upon the analysis of orthotropic plates in which the loads are represented by a harmonic series. Only the first term of the series is used to obtain the coefficients, which are given either in a graphical form [18,21] or in a tabular form [6]. The characterizing parameters α and θ described earlier [Eqs. (1.11) and (1.12), respectively] are used as the basis of the method. The distribution coefficients, which are given for nine standard reference points and load positions across the bridge width, are plotted or tabulated against values of θ. The charts or tables are given for two values of α, namely, $\alpha = 0.0$ and $\alpha = 1.0$. Values of the coefficients K_α for intermediate values of α are obtained by the following interpolation function.

$$K_\alpha = K_0 - (K_0 - K_1)(\alpha^{0.5}) \tag{1.43}$$

where K_0 and K_α are the corresponding coefficients for α equal to 0.0 and 1.0, respectively.

For analysis by this method, the applied loads are converted into equivalent concentrated loads at the standard locations. The coefficients for each load location are then added in order to account for all the equivalent loads in the load case under consideration.

To compensate for possible errors resulting from the representation of loads by only one harmonic, it has been suggested in Ref. 18 that the computed longitudinal moments be increased by an arbitrary 10 percent.

Cusens and Pama [9] have improved the distribution coefficient approach by taking seven terms of the harmonic series and by extending the range of α up to 2.0. Their method also makes use of an interpolation equation similar to Eq. (1.43).

As shown in Ref. 4, interpolation by Eq. (1.43) can introduce substantial errors, especially when the value of θ is small.

Distribution Coefficient Method
Based upon Semicontinuum Idealization

In Ref. 12 extensive use has been made of harmonic analysis as applied to the semicontinuum structure defined earlier. The idealization is a good one for several types of bridges, for example, the slab-on-girder type in which longitudinal flexural rigidity is largely concentrated at the positions of the girders and transverse rigidity is uniformly distributed along the slab. As originally presented, the method did not take into account transverse torsion; this omission was subsequently taken care of [13].

TABLE 1.1   Some AASHTO *D* Values

Bridge type	D, m	
	Bridge designed for one traffic lane	Bridge designed for two or more traffic lanes
Slab-on girder bridge with steel or prestressed girders	2.13	1.67
Slab-on-girder bridge of T-beam construction	1.98	1.83
Slab-on-girder bridge with timber girders	1.83	1.52

The AASHTO Method of Analysis

The AASHTO specifications for highway bridges [1] permit a simplified method for obtaining longitudinal moments and shears due to live loads. According to this method, a longitudinal girder, or a strip of unit width in the case of slabs, is isolated from the rest of the structure and treated as a one-dimensional beam. The beam is subjected to loads comprising one line of wheels of the design vehicle multiplied by a load fraction S/D. S is the girder spacing in the case of slab-on-girder bridges, or 1 unit width in the case of slab bridges; and D, which has the units of length, is specified to have a certain value according to the bridge type. Values of D for various slab-on-girder bridges, as given in the AASHTO specifications [1] are reproduced in Table 1.1. The concept of the quantity D is explained in Sec. 4.1.

The AASHTO D values for this extremely simplified method are based upon research reported in Refs. 19 and 22. The simplicity of the method, however, does take its toll in accuracy. In the AASHTO method, the value of D depends only on the bridge type; however, it is intuitively obvious, and confirmed by accurate analysis, that the load distribution in a narrow and long bridge is different from that in a short and wide bridge of the same type. The AASHTO method is unable to allow for differences in pattern of load distribution arising from such factors as the aspect ratio of the bridge.

Basic Assumptions

The two basic assumptions which are common to both the Guyon-and-Massonnet and AASHTO methods are as follows.

1. For a given bridge the transverse distribution pattern is the same for all load effects, i.e., deflections, moments, and shears.

2. The transverse distribution of load effects is independent of the longitudinal positions of loads and reference sections.

It has been shown earlier in the chapter that the characterizing parameters for different responses can be different. From this it follows that the transverse distributions of different responses can also be different. Using rigorous analyses it can be readily demonstrated that the transverse distribution of responses become "peakier," i.e., more localized, as the number of deflection derivatives to which the response is related increases. For example, as shown in Fig. 1.14, the distribution coefficients for deflection under a concentrated load in a bridge are smaller than those for longitudinal moments, which in turn are smaller than those for longitudinal shear. This follows from the fact that moments are related to the second derivatives of deflection, and shears to the third. It is noted that the distribution coefficient for a certain response is given by the ratio of the actual intensity of the response to the average intensity of the same response.

As discussed in Sec. 5.1, the transverse distribution of a given response is not exactly uniform along the bridge length, despite the fact that assumption 2 above is basic to the two simplified methods. This assumption can be justified on the grounds that the live loading on a bridge comprises a large number of concentrated loads which are well distributed along the span of the bridge. Under such a loading, the pattern of transverse load distribution does not vary significantly along the span.

Assumption 2 is also incorporated in most of the simplified methods given in this book. As demonstrated in Sec. 8.3, slight deviations from

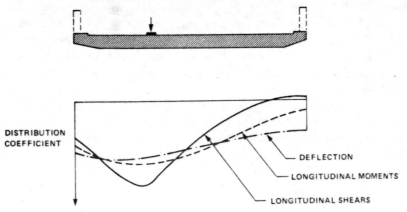

Figure 1.14   Distribution patterns for different responses.

this assumption, which do take place in real life, have an insignificant effect on the accuracy of the results.

References

1. American Association of State Highway and Transportation Officials (AASHTO): *Standard Specifications for Highway Bridges*, Washington, D.C., 1977.
2. Bakht, B., Cheung, M. S., and Aziz, T. S.: The application of a simplified method of calculating longitudinal moments to the proposed Ontario Highway Bridge Design Code, *Canadian Journal of Civil Engineering*, 6(1), 1979, pp. 36–50.
3. Bakht, B., and Jaeger, L. G.: Effect of vehicle eccentricity on longitudinal moments in bridges, *Canadian Journal of Civil Engineering*, 10(4), 1983, pp. 582–599.
4. Bakht, B., and Jaeger, L. G.: Simplified analysis for slab-on-girder bridges, *Bridge and Structural Engineer, India*, 12(4), 1982, pp. 5–30.
5. Bakht, B., Jaeger, L. G., and Cheung, M. S.: Transverse shear in multibeam bridges, *Journal of the Structural Division, ASCE*, 109(4), April 1981, pp. 936–949.
6. Bares, R., and Massonnet, C.: *Analysis of Beams and Grids and Orthotropic Plates by the Guyon-Massonnet-Bares Method*, Crossby Lockwood, London, 1968.
7. Cheung, M. S., and Cheung, Y. K.: Analysis of curved box girder bridges by finite strip method, *Publications, IABSE*, 31-1, 1971, pp. 1–20.
8. Cheung, Y. K.: *Finite Strip Method in Structural Analysis*, Pergamon Press, Oxford, England, 1976.
9. Cusens, A. R., and Pama, R. P.: *Bridge Deck Analysis*, Wiley, London, 1975.
10. Guyon, Y.: Calcul des ponts larges à poutres multiples solidarisées par des entretoises, *Annales des Ponts et Chaussées*, no. 24, 1946, pp. 553–612.
11. Hambly, E. C.: *Bridge Deck Behavior*, Chapman and Hall, London, 1976.
12. Hendry, A. W., and Jaeger, L. G.: *The Analysis of Grid Frameworks and Related Structures*, Prentice-Hall, Englewood Cliffs, N.J., 1958.
13. Jaeger, L. G., and Bakht, B.: *A General Method for the Analysis of Slab-on-Girder Bridge Decks*, Structural Research Report SRR-84-09, Ministry of Transportation and Communications, Downsview, Ontario, Canada, 1984.
14. Jaeger, L. G., and Bakht, B.: The grillage analogy in bridge analysis, *Canadian Journal of Civil Engineering*, 9(2), 1982, pp. 224–235.
15. Massonnet, C.: Méthode de calcul des ponts à poutres multiples tenant compte de leur résistance à la torsion, *Publications, IABSE*, 10, 1950, pp. 147–182.
16. Massonnet, C., and Gandolfi, A.: Some exceptional cases in the theory of multi-girder bridges, *Publications, IABSE*, 27, 1967, pp. 73–94.
17. Ministry of Transportation and Communications: *Ontario Highway Bridge Design Code (OHBDC)*, 2d ed., Downsview, Ontario, Canada, 1983.
18. Morice, P. B., and Little, G.: *The Analysis of Right Bridge Decks Subjected to Abnormal Loading*, Report Db 11, Cement and Concrete Association, London, 1956.
19. Newmark, N. M.: Design of I-beam bridges, Highway Bridge Floor Symposium, *Journal of the Structural Division, ASCE*, 74(STI), March 1948, pp. 305–331.
20. Pama, R. P., and Cusens, A. R.: Edge beam stiffening of multi-beam bridges, *Journal of the Structural Division, ASCE*, 93(ST2), April 1967, pp. 141–161.
21. Rowe, R. E.: *Concrete Bridge Design*, C. R. Books, London, 1962.
22. Sanders, W. W., and Elleby, H. A.: *Distribution of Wheel Loads on Highway Bridges*, National Cooperative Highway Research Program Report 83, Transportation Research Board, Washington, D.C., 1970.
23. Spindel, J. E.: A study of bridge slabs having no transverse flexural stiffness, Ph.D. thesis, London University, London, 1961.
24. Timoshenko, S. P., and Woinowsky-Krieger, S.: *Theory of Plates and Shells*, McGraw-Hill, New York, 1959.
25. Zienkiewicz, O. C.: *The Finite Element Method in Engineering Science*, McGraw-Hill, New York, 1971.

CALCULATION OF CHARACTERIZING PARAMETERS

2.1 Introduction

The first step in analysis by most of the simplified methods given in this book is to calculate the values of the relevant characterizing parameters. Since many of these parameters are used in several of the methods, the procedures for their calculation are given separately in this chapter. It is important for the engineer not only to be able to calculate the values of these characterizing parameters, but also to be aware of the typical ranges of their values so that numerical errors in calculations may be easily detected. Typical ranges of the values of the various parameters are therefore given for different bridge types.

2.2 Torsional Parameter α and Flexural Parameter θ

As shown in Chap. 1, the expressions for α and θ are:

$$\alpha = \frac{D_{xy} + D_{yx} + D_1 + D_2}{2(D_x D_y)^{0.5}} \tag{2.1}$$

$$\theta = \frac{b}{L}\left(\frac{D_x}{D_y}\right)^{0.25} \tag{2.2}$$

where the notation is as shown in Fig. 2.1 and is as follows:

x direction = the longitudinal direction, i.e., the direction of traffic flow

y direction = the transverse direction (perpendicular to the longitudinal direction)

D_x = the longitudinal flexural rigidity (corresponding to EI in a longitudinal beam) per unit width

D_y = the transverse flexural rigidity (corresponding to EI in a transverse beam) per unit length

D_{xy} = the longitudinal torsional rigidity (corresponding to GJ in a longitudinal beam) per unit width

D_{yx} = the transverse torsional rigidity (corresponding to GJ in a transverse beam) per unit length

D_1 = the longitudinal coupling rigidity (which is the contribution of transverse flexural rigidity to longitudinal torsional rigidity through Poisson's ratio) per unit width

D_2 = the transverse coupling rigidity per unit length

It is important to note that the manner of transverse load distribution in the various components of a bridge depends mainly upon the sum of the various torsional rigidities, that is, $D_{xy} + D_{yx} + D_1 + D_2$, rather than on their individual values. Further, the usual experimental determination of plate rigidities will give the sum of the torsional rigidities rather than individual values. For the sake of convenience it is sometimes assumed that D_{xy} is equal to D_{yx}. A feel for the physical nature of the various plate rigidities can be developed by reference to Fig. 2.2. This figure shows some isolated portions of a bridge as well as load effects which correspond to the different plate rigidities.

The coupling rigidities D_1 and D_2 can be regarded as the torsional stiffening of the plate due to the lateral Poisson's ratio effect in a strip that

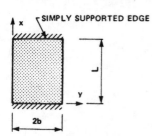

Figure 2.1 Plan of a simply supported bridge of rectangular planform.

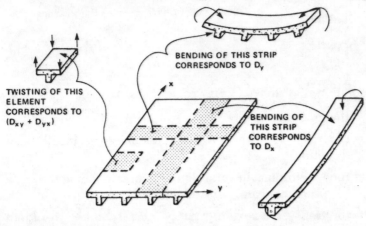

Figure 2.2 Physical significance of the plate rigidities.

is subjected to flexure. An idealized orthotropic plate is of uniform thickness; therefore, D_1 and D_2 are affected by both D_x and D_y. In the case of slab-on-girder bridges the Poisson's ratio effects are restricted to the deck slab only. Therefore, in such bridges it is sufficiently accurate to assume that D_1 and D_2 arise from flexural rigidities which are equal to the flexural rigidity of the deck slab alone, i.e., by D_y in most cases.

Plate rigidities required for the calculation of α and θ for various bridge types can be calculated as shown in the following sections.

Noncomposite Slab-on-Girder Bridges

Steel girder bridges with concrete deck slabs constructed before the 1950s do not have mechanical devices, such as shear studs, to transfer interface shear from the girders to the deck slab. In the absence of any shear transfer at this interface the girders and the deck slab are free to flex about their own neutral axes, as shown in Fig. 2.3. In this case the various plate rigidities can be obtained as follows:

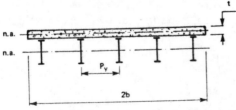

Figure 2.3 Cross section of a noncomposite slab-on-girder bridge.

$$D_x = \frac{E_s I}{P_y} + \frac{E_c t^3}{12(1 - v_c^2)} \approx \frac{E_s I}{P_y}$$

$$D_y = \frac{E_c t^3}{12(1 - v_c^2)} \qquad \approx \frac{E_s t^3}{12n}$$

$$D_{xy} = \frac{G_s J}{P_y} + \frac{G_c t^3}{6} \qquad \approx \frac{G_s\, t^3}{n_s\, 6} \qquad \text{for bridges}$$

$$D_{yx} = \frac{G_c t^3}{6} \qquad\qquad = \frac{G_s\, t^3}{n_s\, 6} \qquad \text{with steel girders} \qquad (2.3)$$

$$D_1 = v_c D_y$$

$$D_2 = v_c D_y$$

where E, G, and v are the modulus of elasticity, shear modulus, and Poisson's ratio, respectively; subscripts s and c refer to steel and concrete, respectively; n is the ratio of the moduli of elasticity of steel and concrete; n_s is the ratio of shear moduli for steel and concrete; I and J are the girder second moments of area and torsion constant, respectively; and other notation is as shown in Fig. 2.3.

Assuming that the Poisson's ratios for steel and concrete are 0.3 and 0.15, respectively, then the ratio of their shear moduli, n_s, is approximately $0.88n$.

The flexural rigidity of the deck slab, bending about its own neutral axis, is usually small compared with the girder rigidities. Therefore, its exclusion from D_x, as reflected by the approximate expression given above, would have a very small effect on α, and an even smaller effect on the various load distributions that are obtained with respect to this parameter. By similar reasoning, the torsional rigidities of steel girders may be excluded from consideration in the calculation of D_{xy}.

Many slab-on-girder bridges without mechanical shear connectors do maintain a certain degree of composite action between the girders and the deck slab. This composite action is developed because of bond and friction between concrete and steel. Nearly complete composite action has been observed in some cases when the top girder flange is partially embedded in the deck slab. While the assumption of noncomposite action may be a safe one in determination of the resistances, it may not be so for the calculation of live-load effects.

A composite girder, being stiffer than the noncomposite one, will attract a greater portion of the live load placed on it. Therefore, live-load analysis ignoring any composite action will predict lower maximum load effects than the girders are likely to attract in real life. It is advisable, in most cases, to ignore the absence of composite action for live-load analysis, and to calculate the characterizing parameters as if the structure were composite.

Composite Slab-on-Girder Bridges with Concrete Deck Slabs

The cross sections of two composite slab-on-girder bridges are shown in Fig. 2.4; Fig. 2.5 shows the cross section of a concrete T-beam bridge.

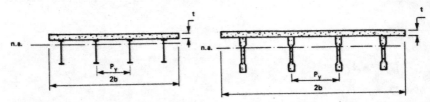

Figure 2.4   Cross sections of composite slab-on-girder bridges with concrete deck slabs.

The plate rigidities of such bridges can be calculated as follows:

$$D_x = \frac{E_g}{P_y} I_g \qquad (2.4a)$$

where I_g is the combined second moment of area of girder and associated portion of deck slab in units of girder material;

$$D_y = \frac{E_c t^3}{12(1 - v_c^2)} \approx \frac{E_c t^3}{12} \qquad (2.4b)$$

$$D_{xy} = \frac{G_g J}{P_y} + G_c \frac{t^3}{6} \qquad (2.4c)$$

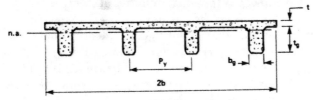

Figure 2.5   Cross section of a T-beam bridge.

$$D_{yx} = G_c \frac{t^3}{6} \qquad (2.4d)$$

$$D_1 = v_c D_y \qquad (2.4e)$$

$$D_2 = v_c D_y \qquad (2.4f)$$

where the subscript g applied to E and G refers to the material of the girder, and other notation is as defined earlier and as shown in Figs. 2.4 and 2.5.

The combined second moment of area of the girder and associated portion of the deck slab is obtained without taking account of the fact that the apparent flexural rigidity of a girder with a very wide flange is reduced owing to the shear lag effects in the flange. It has been shown in Ref. 14 that in most practical cases the reduction of the flexural rigidity of a T beam due to shear lag effects is negligible so far as load distribution is concerned. In other words, small changes in the longitudinal flexural rigidities do not significantly affect the load distribution characteristics of the bridge. Therefore, in most cases it is sufficiently accurate to use the approximate values of flexural rigidities that are obtained by considering full slab widths without taking account of shear lag.

Plate rigidities corresponding to reinforced concrete components are calculated by ignoring the steel reinforcement and by assuming that the concrete is uncracked. It is usual (see, for example, Ref. 16) to ignore the reduction of plate rigidities due to concrete cracking.

The torsional inertia of steel girders, J (also called the torsional constant in handbooks of steel sections), is usually negligible in comparison with the torsional rigidity of the deck slab. Therefore, for bridges with steel girders, the first term of the expression for D_{xy} can be safely neglected. However, torsional inertia of concrete girders, including those in T-beam bridges, can be substantial and should not be ignored.

The torsional inertia of a beam is often calculated by dividing it into a number of rectangles and adding the torsional inertias of the individual rectangles. The torsional inertia of a rectangle of sides w and d, where d is smaller than w, is given by

$$J = Kwd^3 \tag{2.5}$$

where K can be obtained from Fig. 2.6. Thus, the torsional inertia of a girder consisting of three idealized rectangles, as shown in Fig. 2.7, is given by:

$$J = \sum_{n=1}^{3} K_n w_n d_n^3 \tag{2.6}$$

Concrete Slab Bridges

For a solid concrete slab bridge, a partial cross section of which is shown in Fig. 2.8a, D_x and D_y are each equal to $E_c t^3 / 12(1 - v_c^2)$; each of D_{xy} and D_{yx} is equal to $G_c t^3 / 6$; and each of D_1 and D_2 is equal to $vE_c t^3 / 12(1 - v_c^2)$.

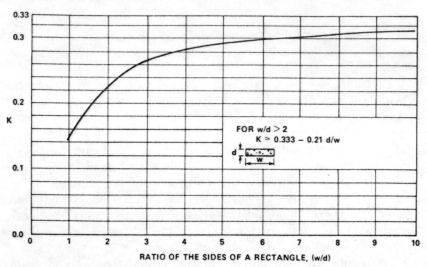

Figure 2.6 Values of the torsion coefficient.

It is noted that the torsional inertia per unit width of a very wide plate is equal to $t^3/3$. Only half this inertia is being considered in the expressions for D_{xy} and D_{yx} because this torsional inertia is equally shared in rigidities, corresponding to the two directions. The halving of rigidity in this manner should be done only for those components which are continuous in both directions, such as deck slabs. For isotropic materials, E_c and G_c are related according to the following relationship:

$$G_c = \frac{E_c}{2(1 + \nu_c)} \tag{2.7}$$

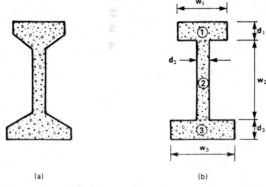

(a)   (b)

Figure 2.7 Idealization for calculating torsional inertia. *(a)* Actual cross section. *(b)* Idealized cross section.

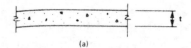

(a)

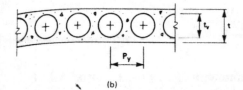

(b)

Figure 2.8 Partial cross sections of slab and voided slab bridges: *(a)* slab; *(b)* voided slab.

Substituting the above plate rigidities and the expression for G_c in Eqs. (2.1) and (2.2), the following simplified expressions are obtained for α and θ of concrete slab bridges.

$$\alpha = 1.0$$

$$\theta = \frac{b}{L} \tag{2.8}$$

The above expressions can be used whether the slab is of reinforced or prestressed concrete construction.

Voided Slab Bridges

It has been shown in Ref. 5 that the plate rigidities of a voided slab bridge (Fig. 2.8b) having centrally placed circular voids can be obtained by the following simplified expressions:

$$D_x = E_c\left(\frac{t^3}{12} - \frac{\pi t_v^4}{64P_y}\right)$$

$$D_y = \frac{E_c t^3}{12}\left[1 - 0.95\left(\frac{t_v}{t}\right)^4\right]$$

$$D_{xy} = \frac{G_c t^3}{6}\left[1 - 0.84\left(\frac{t_v}{t}\right)^4\right] \tag{2.9}$$

$$D_{yx} = D_{xy}$$

$$D_1 = v_c D_y$$

$$D_2 = v_c D_y$$

The above expressions for D_y and D_{xy} are due to Elliot [10].

In the absence of more accurate methods, Eq. (2.9) may also be used for those voided slab bridges in which the circular voids are not symmetrically placed between the top and bottom surfaces.

Waffle Slab Bridges

A waffle slab bridge, a segment of which is shown in Fig. 2.9, may be regarded as a T-beam bridge with frequently spaced diaphragms. For the purpose of simplified analysis, it is sufficiently accurate (as shown in Ref. 2) to disregard the enhancement of longitudinal rigidities due to the presence of transverse ribs and vice versa. Thus:

$$D_x = \frac{E_c}{P_y} I_x \tag{2.10a}$$

where I_x is the combined second moment of area of a longitudinal rib and the associated portion of the deck slab;

$$D_y = \frac{E_c}{P_x} I_y \tag{2.10b}$$

where I_y is the combined second moment of area of a transverse rib and the associated portion of the deck slab;

$$D_{xy} = \frac{G_c J_L}{P_y} + G_c \frac{t^3}{6} \tag{2.10c}$$

$$D_{yx} = \frac{G_c J_T}{P_x} + G_c \frac{t^3}{6} \tag{2.10d}$$

$$D_1 = v_c \times (\text{smaller of } D_x \text{ and } D_y) \tag{2.10e}$$

$$D_2 = D_1 \tag{2.10f}$$

where J_L and J_T are torsional inertias of longitudinal and transverse ribs, respectively; their values can be obtained by using Eq. (2.5).

For refined analysis of waffle slab bridges with complex planforms, it may be prudent to obtain a better estimate of the plate rigidities. For such cases it is recommended to use flexural and coupling rigidity expressions for uncracked sections as given in Ref. 15 and the method of Ref. 13 for torsional rigidities. The latter method accounts for the fact

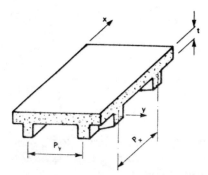

Figure 2.9 A waffle slab segment.

that the torsional inertia of a section composed of two or more rectangles is more than the sum of the individual torsional inertias of the various rectangles [Eq. (2.6)]. The enhancement of the total torsional inertia is caused by the junction effect.

Cellular Structures

Transverse and longitudinal sections of a typical concrete cellular structure are shown in Fig. 2.10. Following the notation given in this figure, the various plate rigidities for a concrete cellular structure having top and bottom flanges of equal thickness can be approximately determined as follows [5, 6]:

$$D_x = 0.5 E_c t_1 H^2$$
$$D_y = 0.5 E_c t_1 H^2$$
$$D_{xy} = G_c t_1 H^2$$
$$D_{yx} = D_{xy} \tag{2.11}$$
$$D_1 = v_c D_y$$
$$D_2 = D_1$$

By substituting the above plate rigidities in Eqs. (2.1) and (2.2), it can be demonstrated that α and θ for concrete cellular structures are approximately given by the following equations:

$$\alpha = 1.0$$
$$\theta = \frac{b}{L} \tag{2.12}$$

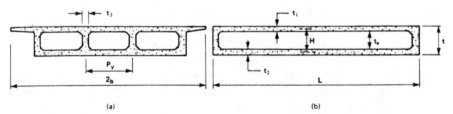

Figure 2.10   (a) Cross section and (b) longitudinal section of a cellular structure.

Properties of Wood

The lack of homogeneity makes wood perhaps the most complex structural material. In setting up a reasonable mathematical model it is assumed that the macrostructure of wood is linearly elastic and that it

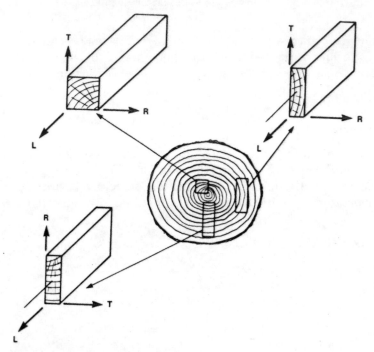

Figure 2.11   Orientation of reference axes.

possesses rhombic symmetry, with the orthogonal axes corresponding to the longitudinal direction of the tree trunk and with the radial and tangential directions of the growth ring pattern of the cross section as shown in Fig. 2.11. This leads to three moduli of elasticity, three shear moduli, and three Poisson's ratios.

The most significant modulus of elasticity is E_L, which corresponds to the longitudinal axis of the tree trunk. The other two moduli of elasticity, that is, E_T and E_R, are both very small compared with E_L, being of the order of one-twentieth of E_L. For bridge analysis, these two moduli can be grouped together. The shear modulus G_{RT}, which is never used in simplified bridge analysis, is extremely small, being of the order of one one-hundredth of E_L. The other two shear moduli, G_{LR} and G_{LT}, are also small, but their values are about one-fifteenth of E_L. These two shear moduli can be grouped together for the purpose of analysis. It is important to note that since wood is not an isotropic material, its moduli of elasticity and shear moduli are not related by an equation of the form of Eq. (2.7). Further details about the properties of wood can be found in Ref. 11.

The properties of wood are subject to fairly wide variations. However, as suggested in Ref. 3, a deterministic analysis of timber bridges requires the mean values of the various parameters. These mean values do differ from species to species. In the absence of actual mean material properties, the following value of E_L can be employed.

$$E_L = 9600 \text{ MPa } (1.39 \times 10^6 \text{ lb/in}^2)$$

Values of E_T and G_{LT} can be obtained from the following approximate relationships.

$$E_T = 0.05 E_L$$
$$G_{LT} \doteqdot 0.07 E_L \qquad\qquad (2.13)$$

Transverse Laminated-Wood Decks on Longitudinal Girders

A typical bridge with a transverse laminated-wood deck on longitudinal girders is shown in Fig. 2.12. The girders may be of sawn timber, glue-laminated wood (glulam), or steel. The transverse decking consists of wood laminates which are nailed together. An important feature of this type of bridge is the orientation of wood in the decking with respect to

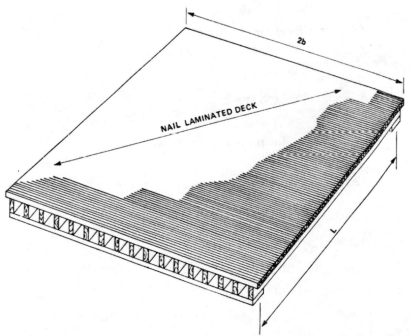

Figure 2.12 A bridge with transverse laminated-wood deck on longitudinal girders.

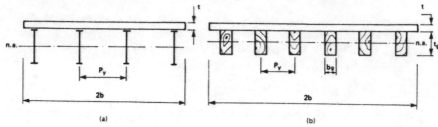

Figure 2.13 Cross sections of bridges with transverse nail-laminated timber deck-ing. (a) Steel girders. (b) Timber girders.

the longitudinal direction. Because of this orientation the wood modulus of elasticity, which, in the event of a composite interaction, can enhance the longitudinal flexural rigidity of the girders, is E_T. Since this modulus is very small, it can be assumed for all practical purposes that the girders flex about their own axes even if there are competent shear transfer devices between the decking and girders.

The cross sections of two of these bridges are shown in Fig. 2.13. The plate rigidities of such bridges can be calculated as follows, using the notation shown.

$$D_x = \frac{E_L}{P_y} \frac{b_g t_g^3}{12} \qquad \text{for timber girders type} \qquad (2.14a)$$

$$= \frac{E_s}{P_y} I \qquad \text{for steel girders type} \qquad (2.14b)$$

where I = second moment of area of a girder;

$$D_y = E_L \frac{t^3}{12} \qquad \text{for both types} \qquad (2.14c)$$

$$D_{xy} = \frac{G_{LT} K t_g b_g^3}{P_y} \qquad \text{for timber girders type} \qquad (2.14d)$$

$$= \frac{G_s}{P_y} J \qquad \text{for steel girders type} \qquad (2.14e)$$

where J = torsional inertia of a girder;

$$D_{yx} = 0.0 \qquad \text{for both types} \qquad (2.14f)$$

$$D_1 = 0.0 \qquad \text{for both types} \qquad (2.14g)$$

$$D_2 = 0.0 \qquad \text{for both types} \qquad (2.14h)$$

K in the expression for D_{xy} can be obtained from Fig. 2.6 for the appropri-ate ratio of t_g/b_g.

Since the individual decking laminates are usually free to rotate, the torsional inertia per laminate is equal to Ktb_L^3, where b_L is the width of the individual laminate. Hence the torsional inertia per unit length of the decking is Ktb_L^2. Since b_L is usually small, being of the order of 40 mm, the torsional inertia becomes negligibly small, and one may take $D_{yx} = 0$.

It is noted that when the individual laminates are prevented from rotating individually, as in the case of glue-laminated or prestressed decking, D_{xy}, D_{yx} are given by the following expressions:

$$D_{xy} = \frac{G_{LT}Kt_g b_g^3}{P_y} + G_{LT}\frac{t^3}{6} \qquad \text{for timber girders type} \qquad (2.15a)$$

$$D_{xy} = \frac{G_s J}{P_y} + G_{LT}\frac{t^3}{6} \qquad \text{for steel girders type} \qquad (2.15b)$$

where J is the torsional inertia of the girder;

$$D_{yx} = G_{LT}\frac{t^3}{6} \qquad \text{for both types} \qquad (2.15c)$$

Prestressed Laminated-Wood Bridges

This new type of bridge, which was developed in Ontario to rehabilitate existing nail-laminated timber bridges (see Refs. 9 and 17) consists of longitudinal vertical wood laminates which are transverse-post-tensioned. A minimum average prestressing pressure of 0.35 MPa (50 lb/in²) is required for the system of laminates to be considered as an integral unit [16]. The prestressing system can be an external one as shown in Fig. 2.14, or one in which the prestressing wires or rods pass

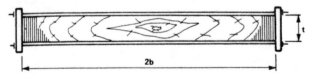

Figure 2.14 Cross section of a transversely post-tensioned, longitudinally laminated-wood bridge.

through holes drilled midway between the top and bottom surfaces. The laminates do not run continuously along the length of the bridges. However, butt joints in laminates are restricted to only one per four laminates in any 1 m (3.3 ft) of bridge length. For a bridge complying with the above restrictions, the effective plate rigidities are given by the following equations.

$$D_x = \frac{E_L t^3}{12}$$

$$D_y = \frac{E_t t^3}{12}$$

$$D_{xy} = G_{LT}\frac{t^3}{6} \qquad\qquad (2.16)$$

$$D_{yx} = D_{xy}$$

$$D_1 = 0.0$$

$$D_2 = 0.0$$

The above expressions also apply to a glue-laminated wood bridge.

Wood-Concrete Composite Bridges

Wood-concrete composite bridges consist of vertical wood laminates with an overlay of concrete as shown in Fig. 2.15. The laminates are held together by nails. The composite action between wood and concrete is made possible by a variety of devices. The Ontario highway bridge design code [16] permits the use of the two interface arrangements shown in Fig. 2.16. Plate properties for bridges in which full composite action can be justifiably assumed can be calculated as follows:

$$D_x = E_c I_c \qquad\qquad (2.17a)$$

where I_c is the combined second moment of area, in concrete units, of a unit width of the bridge;

$$D_y = E_c\frac{t^3}{12} \qquad\qquad (2.17b)$$

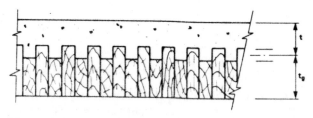

Figure 2.15 Partial cross section of a wood-concrete composite bridge.

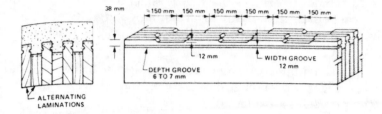

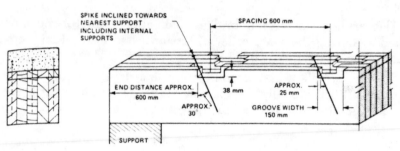

Figure 2.16 Ontario-approved wood-concrete interfaces for full composite action.

$$D_{xy} = G_c \frac{t^3}{6} \tag{2.17c}$$

$$D_{yx} = D_{xy} \tag{2.17d}$$

$$D_1 = vD_y \tag{2.17e}$$

$$D_2 = D_1 \tag{2.17f}$$

where t is obtained as shown in Fig. 2.15 in case of laminates of alternating depths. When all laminates are of the same depth, t is equal to the depth of the concrete overlay.

Typical α and θ Values

Torsional parameter α It has been shown (e.g., in Ref. 4) that values of α for different bridge types fall into distinct and separate ranges.

The parameter α can be regarded as a measure of the ratio of representative torsional rigidity to representative flexural rigidity. Typical ranges for α are as follows:

1. Bridges incorporating wood beams and laminates (but excluding wood-concrete composite bridges) have very small torsional rigidities D_{xy} and D_{yx} because of the inherently small modulus of rigidity of wood. Accordingly, α values for these bridges are low, ranging from nearly zero to about 0.02.

2. The torsional rigidities of wood-concrete composite bridges and of slab-on-girder bridges with steel girders are contributed mainly by the deck slabs; their α values are about 0.1.

3. Slab-on-girder bridges with concrete girders are a little stiffer torsionally than wood-concrete composite, and therefore their α values are a little larger. Typical values of α range between 0.1 and 0.2 for this type of bridge.

4. Slab bridges, voided slab bridges, and cellular structures can be regarded as having the same order of torsional and flexural stiffness. Hence, α is approximately 1.0 for these structures.

5. The use of hollow beams, as in multispine structures, makes the bridge more stiff torsionally than it is in flexure. Therefore, for these bridges α is greater than 1.0, usually ranging between 1.5 and 2.0.

Flexural parameter θ The parameter θ can be regarded as a measure of the ratio of deflection stiffnesses due to flexure in the longitudinal and transverse directions. Hence wide and short bridges have higher values of θ than long and narrow ones. For most practical cases θ ranges between 0.25 and 2.0, the lower figure being for very long and narrow bridges and the upper figure for very wide and short ones. The concept of the (α, θ) space, as shown in Fig. 2.17, may be found useful in the rapid estimation of the values of α and θ for bridges. The figure shows the various zones of the space into which the α and θ values of various types of bridges are likely to fall.

It should be noted that the zones of typical values of α are for usual bridges and should not be taken to be rigid boundaries. There will sometimes be cases of bridges with unusual cross sections for which the values will fall outside the ranges shown. For example, a slab-on-girder bridge with an unusually thick deck slab may have α greater than 0.2. One can imagine a series of slab-on-girder bridges in which the deck slab grew steadily thicker while the size of the girders steadily diminished. Clearly the behavior would steadily approach that of a solid slab bridge for which the value of α is close to 1.0. The "limiting" value of $\alpha = 0.2$ shown in Fig. 2.17 for slab-on-girder bridges is a representative upper limit for practical cases of this bridge type.

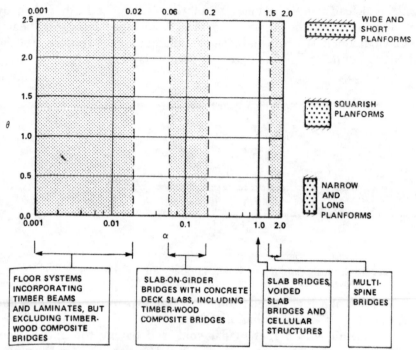

Figure 2.17 Typical α and θ values for various bridge types.

2.3   Articulated Plate Parameter β

As shown in Chap. 1, the expression for β is

$$\beta = \pi \left(\frac{2b}{L} \right) \left(\frac{D_x}{D_{xy}} \right)^{0.5} \tag{2.18}$$

where the notation is as defined in Sec. 2.2 and in Fig. 2.1.

Unlike slab bridges, in which D_{xy} represents half the torsional inertia of the slab and D_{yx} the other half, D_{xy} in Eq. (2.18) refers to the full torsional stiffness of the prismatic beams. Therefore, expressions for D_{xy} for this case may be different from those given in Sec. 2.2. Expressions for D_x and D_{xy} are given below for a number of bridge types, following which representative values of the parameter β are given.

Multibeam Bridges

Multibeam bridges consist of precast concrete longitudinal beams placed side by side and connected to adjacent beams in such a manner that the transfer of load from one beam to another is achieved other than by transverse flexure. The cross section of such a bridge having solid beams is shown in Fig. 2.18. For this bridge the plate rigidities required for the calculation of β can be obtained by the following expressions:

$$D_x = \frac{E_c t^3}{12}$$

$$D_{xy} = G_c K t^3 \qquad \text{if } P_y > t \qquad\qquad (2.19)$$

$$D_{xy} = G_c K P_y^2 t \qquad \text{if } P_y < t$$

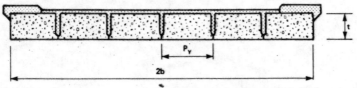

Figure 2.18   Cross section of a multibeam bridge with solid beams.

The value of K is obtained from Fig. 2.6 and corresponds to the value of P_y/t, or t/P_y if P_y is less than t.

For a beam with a rectangular void, and having top and bottom flanges of equal thickness, as shown in Fig. 2.19a, the values of D_x and D_{xy} are approximately given by:

$$D_x = 0.5 E_c t_1 H^2$$

$$D_{xy} = G_c \frac{J}{P_y} \qquad\qquad (2.20)$$

where J is the torsional inertia of one beam and is given by:

$$J = \frac{4A^2}{\oint ds/t'} \qquad\qquad (2.21)$$

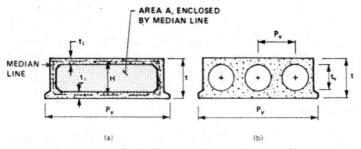

Figure 2.19   Beams of multibeam bridges: (a) beam with a rectangular void; (b) beam with circular voids.

where A is the area enclosed by the median line of the closed section of the beam as shown in Fig. 2.19a; and $\oint ds/t'$ refers to the contour integral along the median line of the reciprocal of the wall thickness.

The plate rigidities of multibeam bridges having beams with circular voids which are symmetrically placed between the top and bottom surfaces of the bridge (Fig. 2.19b) can be obtained by the following expressions:

$$D_x = E_c \left(\frac{t^3}{12} - \frac{\pi t_v^4}{64 P_v} \right)$$

$$D_{xy} = G_c \frac{t^3}{3} \left[1 - 0.84 \left(\frac{t_v}{t} \right)^4 \right]$$

(2.22)

The above approximate expression for D_{xy} for the voided slab given is derived in Ref. 10.

Multispine Bridges

Bridges consisting of beams of closed cross sections and a concrete deck on top are referred to as *multispine bridges*. The cross section of such a bridge is shown in Fig. 2.20. The beams may be of precast concrete construction or of steel. Plate rigidities for this type of bridge can be obtained by the following expressions:

$$D_x = \frac{E_c}{P_y} I_g$$

(2.23a)

where I_g is the combined second moment of area of a spine and associated deck slab, in units of deck slab concrete;

$$D_{xy} = \frac{G_c}{P_y} \left(\frac{4A^2}{\oint ds/n_t t'} \right)$$

(2.23b)

where A is the area enclosed by the median line passing through the walls of the closed cross section, as shown in Fig. 2.20. Portions of the deck slab

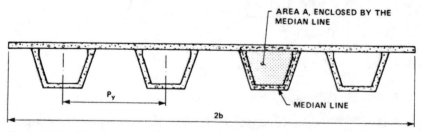

Figure 2.20   Cross section of a multispine bridge.

which do not form the closed section, and appendages to the girder walls (such as stiffeners to steel plates), have negligible influence on the torsional rigidity. Therefore these components can safely be excluded from consideration in the calculation of D_{xy}.

In the case of all concrete construction n_s is equal to 1.0, and for steel-concrete construction it is equal to $0.88n$.

Nail-Laminated Timber Bridges

A bridge consisting of longitudinal wood laminates which are joined together by means of nails can be idealized as an articulated plate because the transverse load distribution in this sort of structure takes place only through shear. However, this idealization is limited to new bridges which have not been subjected to a large cycle of live loads. Field and laboratory measurements by several research workers (see, for example, Refs. 9 and 17) have confirmed that nail connections in these bridges suffer significant deterioration under repeated loads. When nails become loose in their holes after a few years of service, a loaded laminate deflects freely a little before engaging adjacent laminates. Thus the representation of the system of laminates as an articulated plate ceases to be valid after a few years of service. The following plate rigidities are therefore valid only for new bridges.

$$D_x = E_L \frac{t^3}{12}$$

$$(2.24)$$

$$D_{xy} = G_{LT} K t b_g^2$$

where E_L and G_{LT} are as defined in Sec. 2.2; b_g is the width of the laminate, as shown in Fig. 2.21; and K is obtained from Fig. 2.6 for the ratio t/b_g.

In the expression for D_{xy} it is assumed that t is larger than b_g; if it is not, then the right-hand side of the equation should be replaced by $G_{LT} K t^3$.

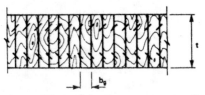

Figure 2.21 Partial cross section of a nail-laminated wood bridge.

Prestressed Laminated-Wood Bridges

In a prestressed wood bridge (see Fig. 2.14), the transverse prestressing enables the bridge to develop transverse flexural rigidity. Thus, longitudinal moments and shears can be calculated by simplified methods derived from general orthotropic plate theory, i.e., by the (α, θ) method. For determining transverse shear by simple methods, however, it is necessary to idealize these bridges as articulated plates. It is noted that in these bridges transverse flexural rigidity D_y is about one-twentieth of the longitudinal rigidity. It has been shown that when D_y is so small compared with D_x, then ignoring it altogether has relatively little effect on load distribution, and that the relatively small error involved is on the safe side so far as transverse shear is concerned. For prestressed laminated-wood bridges the plate rigidities for determining β may be taken as:

$$D_x = E_L \frac{t^3}{12}$$

$$D_{xy} = G_{LT} \frac{t^3}{3}$$

(2.25)

The difference in expressions for D_{xy} as given above and as calculated for determining α and θ [Eq. (2.16)] should be noted. In Eq. (2.16) the torsional rigidity is divided equally between D_{xy} and D_{yx}. The D_{xy} equation, (2.25), is based on the assumption that D_{yx} is equal to zero. Thus all the torsional rigidity is assigned to D_{xy}. As noted in Sec. 2.2, load distribution in an orthotropic plate depends mainly upon the total torsional rigidity, regardless of the manner in which this total rigidity is distributed between the longitudinal and transverse directions.

Typical β Values

It has been noted earlier that α and θ provide measures of torsional to flexural stiffness and longitudinal to transverse flexural deflection stiffness, respectively. The parameter β includes aspects of both these measures. Hence it is not surprising that it is not possible to group different bridge types into separate ranges of β. However, it is still readily possible to estimate the value of β if Eq. (2.18) is rewritten in the following form:

$$\beta = \pi \left(\frac{2b}{L} \right) C$$

(2.26)

where C, which is equal to $(D_x/D_{xy})^{0.5}$ and which provides a measure of relative torsional stiffness of the bridge, can be grouped into different ranges for different bridge types. Values of C typically range between 0.8 and 2.5 for multibeam and multispine bridges. Bridges with beams or spines that are wider than they are deep fall at the lower end of the range of values of C, and bridges with deeper but narrow beams or spines have larger values of C. This is shown graphically in Fig. 2.22.

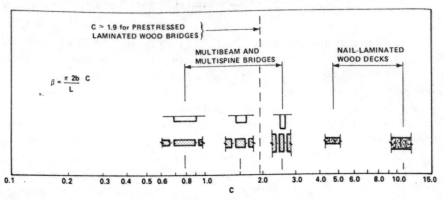

Figure 2.22 Typical values of C.

Typical nail-laminated timber bridges have values of C ranging between 4.7 and 11.7. Deeper laminates have higher values of C than shallower ones.

By using Eq. (2.25) the expression for C for prestressed wood bridges is reduced to:

$$C = \left(\frac{0.25E_L}{G_{LT}}\right)^{0.5} \tag{2.27}$$

If G_{LT} is assumed to be equal to $0.07E_L$, as recommended in Sec. 2.2, then C for prestressed wood bridges is approximately equal to 1.9.

From Fig. 2.22 the value of C can be easily picked up for a given bridge, and using Eq. (2.26) the value of β can be approximately calculated.

2.4 Shear-Weak Plate Parameter δ

As shown in Chap. 1, the expression for the shear-weak plate parameter δ is:

$$\delta = \pi^2 \frac{b}{L^2}\left(\frac{D_x}{S_y}\right)^{0.5} \tag{2.28}$$

where S_y is the transverse shear rigidity per unit length and other notation is as defined in Sec. 2.2. S_y is equal to the product of the shear modulus and the equivalent transverse shear area per unit length.

The derivation of D_x and S_y per unit length for a variety of bridge types is discussed below, and typical values of δ are then given.

Cellular Structures

For a cellular structure D_x is obtained from Eqs. (2.11). Following the notation of Fig. 2.10, the transverse shear rigidity of a concrete cellular structure is given by:

$$S_y = \frac{E_c(t_1^3 + t_2^3)}{P_y^2}\left[\frac{t_3^3 P_y}{t_3^3 P_y + H(t_1^3 + t_2^3)}\right] \tag{2.29}$$

The above equation, which is from Ref. 7, reduces to the following when the top and bottom flanges have the same thickness t_1:

$$S_y = \frac{2E_c t_1^3}{P_y^2[1 + 2(H/P_y)(t_1/t_3)^3]} \tag{2.30}$$

Alternatively, S_y can be approximately obtained from the following simple expression:

$$S_y = 0.5 E_c F_1 t \tag{2.31}$$

where F_1 can be read directly from charts given in Fig. 2.23. Linear interpolation should be used for obtaining intermediate values.

Voided Slab Bridges

D_x for concrete voided slab bridges with circular voids can be obtained from Eqs. (2.9); the value of S_y can be obtained from the following expression:

$$S_y = G_c F_2 t \tag{2.32}$$

where F_2, corresponding to the values of t_o/P_y and t_o/t, can be read directly from Fig. 2.24, which is adapted from Ref. 8. Linear interpolation should be used for obtaining intermediate values.

Equivalent transverse shear area and flexural rigidity are sometimes obtained by replacing circular voids with rectangular voids of equivalent area. It has been demonstrated in Ref. 5 that this practice leads to significant underestimation of both the shear area and the flexural rigidity. It is recommended that the practice of replacing circular voids with rectangular ones not be used.

Transverse Laminated-Wood Decks on Longitudinal Girders

Because of the low modulus of rigidity of wood, transverse wood decking in bridges with steel or wood girders (Fig. 2.12) has smaller shear rigidity

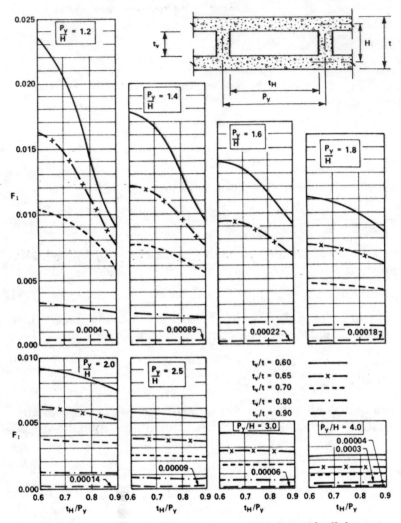

Figure 2.23 Charts for equivalent transverse shear area of cellular structures.

than concrete decking. In most cases this small shear rigidity has a negligible effect on load distribution in the bridge. However, it may be prudent to consider its effects in wide and short bridges having widely spaced steel girders. For such bridges, D_x, of course, can be obtained from Eq. (2.14), and the value of S_y can be obtained from the following:

$$S_y = G_{LT}t \qquad (2.33)$$

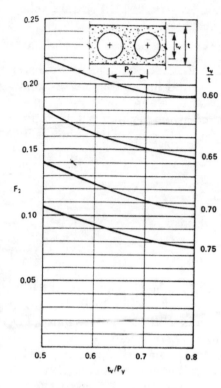

Figure 2.24 Chart for equivalent transverse shear area of cellular structures.

Typical δ Values

Typical δ values are as follows (see Fig. 2.25):

1. The smallest value of δ corresponds to a bridge with no voids, i.e., to a solid slab bridge. For this bridge type δ can have a value as small as 0.005 and as large as 0.1.

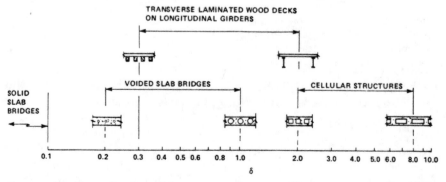

Figure 2.25 Typical values of δ.

2. For voided slab bridges, transverse shear stiffness is intermediate in value between solid bridges and cellular structures. Typical δ values for these bridges range between 0.2 and 1.0.

3. Cellular structures have δ values ranging between 2.0 and 8.0.

As can be seen from the charts of Fig. 2.23, the equivalent shear area is very sensitive to the spacing of webs and flange thicknesses. Short bridges with bigger voids have the larger δ values.

For bridges with transverse laminated-wood decking on longitudinal girders, δ can range between 0.3 and 2.0, with most of the values falling between 0.3 and 0.8.

2.5   Edge-Stiffening Parameter λ

As shown in Chap. 1, the expression for λ is:

$$\lambda = \frac{EI}{L}\left(\frac{1}{D_x^3 D_y}\right)^{0.25} = \frac{EI}{LD_x}\left(\frac{D_x}{D_y}\right)^{0.25} \tag{2.34}$$

where I is the second moment of area of a conceptual beam which, as shown in Fig. 2.26, is symmetrically placed relative to the idealized orthotropic plate. I is calculated in the same material units as those of D_x and D_y, for example, in units of deck slab concrete, and E corresponds to the relevant material units. The conceptual beam may represent physical edge stiffening, such as a barrier wall or raised curb for pedestrians; or it may represent the outer part of the cross section with respect to the outermost line of wheels of a vehicle, shown in Fig. 2.27. I may also be a combination of a physical edge beam and vehicle-edge distance. Plate rigidities D_x and D_y for different bridge types can be calculated according

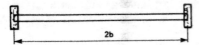

2b

Figure 2.26 Cross section of an idealized orthotropic plate with conceptual symmetrical edge beams.

to the procedures given in Sec. 2.2. Here the calculation of EI, for the sake of convenience, is divided into two parts as follows:

$$EI = EI_b + EI_e \tag{2.35}$$

where I_b refers to the second moment of area provided by a physical edge beam and I_e to the vehicle-edge portion of the cross section. In the following, separate procedures are given for calculating EI_b and EI_e.

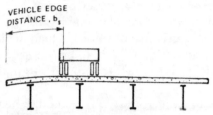

Figure 2.27 Illustration of vehicle-edge distance.

Edge Beams in Slab Bridges and Voided Slab Bridges

Typical edge-stiffening situations are illustrated in Fig. 2.28. It can readily be appreciated that because of the presence of an outstand, the neutral axis of the slab "curls up" near the edges. If the bridge is very wide in comparison with its span, then the moving up of the neutral axis will be limited to the outer portions. On the other hand, if the bridge is very long, then the whole cross section may flex about a common, almost straight, neutral axis. In this case the whole bridge would flex like a U section. Clearly, the span length has an influence on the width of the portion of the slab that may be regarded as acting compositely with the outstand about a common neutral axis.

Following the advice given in Ref. 12, it is recommended that a maximum slab width equal to $L/6$ be considered to act compositely with the outstand. This width should not be greater than b, or greater than $b_c + 0.6$ m (2 ft), where b_c is the width of the edge of the curb or barrier wall as shown in Fig. 2.28. The latter restriction ensures that the portion

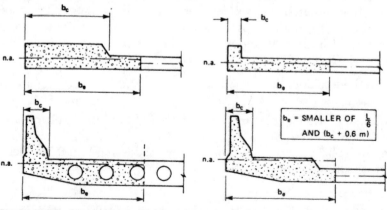

Figure 2.28 Portions of slab bridge cross sections which are considered integral with conceptual edge beams.

of the slab inside the outer line of wheels of the design vehicle is not considered as part of the conceptual edge beam. As described in Chap. 3, the simplified methods for determining longitudinal moments and shears are developed for the most eccentric transverse positions of load, which require the outer line of wheels of the design vehicle to be 0.6 m (2 ft) away from the edge of the curb; this leads to the term 0.6 m (2 ft) in the above expression.

If the combined second moment of area of the physical edge beam and the portion of slab, or voided slab, of width specified above (as shown shaded in Fig. 2.28), is designated as I_B, then EI_b is given by:

$$EI_b = EI_B - b_e D_x \qquad (2.36)$$

where b_e is the width of slab considered in I_B. The second term on the right-hand side of the equation discounts the contribution of the slab flexing about its own neutral axis. The slab flexural rigidity is considered in analysis by means of α and θ.

It is noted that concrete barrier walls are usually made discontinuous along the span. The discontinuity, which may consist of a gap or merely a cold joint, certainly does reduce the effective flexural rigidity of the barrier wall. However, the reduction is rarely more than about 20 percent. It is recommended that, as a conservative approach, the flexural rigidity of a discontinuous barrier wall be reduced by 20 percent.

Edge Beams in Bridges with Girders and Bridges with Webs

Various edge-stiffening conditions in composite slab-on-girder bridges and cellular structures are shown in Fig. 2.29. The presence of outstands on both sides of the deck slabs makes the determination of the flexural rigidity of a conceptual edge beam more complex, and therefore more tentative, than in slab bridges. However, as an approximation, the upper limit of the width of the cross section that can be considered to act compositely with the outstand may be taken as $L/8$. The procedure for determining EI_b is as follows:

1. Find the smaller of $L/8$ and $b_c + 0.60$ m (2.0 ft) and designate this quantity as b_e.

2. Calculate the second moment of area of a width b_e of the cross section measured transversely from the free edge of the bridge. This portion of the bridge cross section is shown shaded in Fig. 2.29. Designate this second moment of area as I_B.

3. Calculate EI_b, using Eq. (2.36).

It is recommended that, as with slab bridges, contributions by discon-

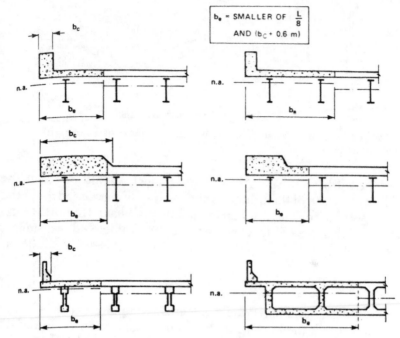

b_e = SMALLER OF $\dfrac{L}{8}$

AND $(b_C + 0.6 \text{ m})$

Figure 2.29 Portions of bridges with girders and webs which are considered integral with conceptual edge beams.

tinuous barrier walls to the flexural stiffness of the conceptual edge beam should be reduced by about 20 percent.

For long and narrow bridges follow the procedure given in Sec. 9.2 and illustrated in Fig. 9.5.

Vehicle-Edge Distance

The vehicle-edge distance b_e depends upon the width of the curb and the minimum distance specified in the code between the face of the curb and the center line of the outer line of wheels. The distance is specified by AASHTO [1] to be 2.0 ft (0.61 m) for all cases. The Ontario code [16] specifies this distance to be 0.60 m (1.97 ft) for the ultimate limit state; however, for the serviceability limit state of fatigue this distance relates to the normal lateral traveling position of vehicles in the outside lanes. Both the AASHTO and OHBDC simplified methods are developed by considering vehicles in the most eccentric positions to be at a distance of 1.0 m (3.28 ft) from the longitudinal free edge. This includes the above-mentioned 0.60 m (1.97 ft) and a curb width of about 0.40 m (1.31 ft).

The conceptual edge beam, which represents an increased vehicle-edge distance (due either to a larger curb width or to a greater distance

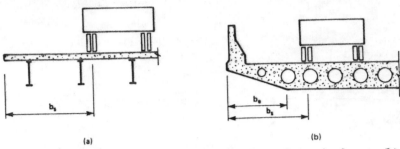

(a)   (b)

Figure 2.30 Vehicle-edge distance b_s. (a) Bridge without edge beams. (b) Bridge with edge beams.

between the curb and the outer line of wheels), should discount the 1.0-m (3.28-ft) width of the bridge, which is already accounted for in the simplified methods for longitudinal moments and shears. Hence the flexural rigidity of the conceptual beam representing vehicle-edge distance, that is, EI_e in bridges without physical edge stiffening (see Fig. 2.30a), is given by

$$EI_e = (b_s - 1.0)D_x \tag{2.37}$$

where b_s is in meters and other quantities are in compatible units. If other units are to be used, then the quantity 1.0 in the above equation should be converted from meters to appropriate units. D_x is calculated as given in Sec. 2.2 for the appropriate bridge type.

In the case of bridges with physical edge beams, a certain width of the bridge, that is, b_c, is already taken into account in the calculation of I_B. Therefore, the expression for I_e should discount that width. Hence, for bridges with physical edge beams (see Fig. 2.30b):

$$EI_e = (b_s - b_c - 1.0)D_x \tag{2.38}$$

As mentioned earlier, the total flexural rigidity of the conceptual edge beam, accounting for both physical edge beams and increased vehicle-edge distance, is given by Eq. (2.35). Again the quantity 1.0 in the above expression is applicable when b_s and b_e are in meters. When b_s and b_e are in feet, the following expression should be used:

$$EI_e = (b_s - b_e - 3.28)D_x \tag{2.39}$$

When the quantity within the parentheses is negative, then EI_e should be assumed to be zero.

References

1. American Association of State Highway and Transportation Officials (AASHTO): *Standard Specifications for Highway Bridges*, Washington, D.C., 1977.

2. Bakht. B.: Discussion on rigidities of concrete waffle-type slab bridges, *Canadian Journal of Civil Engineering*, 7(1), 1980, pp. 198 – 200.

3. Bakht. B.: Statistical analysis of timber bridges, *Journal of the Structural Division, ASCE*. 109(8), August 1983, pp. 1761 – 1779.

4. Bakht. B., Cheung, M. S., and Aziz, T. S.: Application of a simplified method of calculating longitudinal moments to the proposed Ontario Highway Bridge Design Code, *Canadian Journal of Civil Engineering*. 6(1), 1979, pp. 36 – 50.

5. Bakht, B., Jaeger, L. G., and Cheung, M. S.: Cellular and voided slab bridges, *Journal of the Structural Division, ASCE*, 107(ST9), September 1981, pp. 1797 – 1813.

6. Bakht, B., Jaeger, L. G., Cheung, M. S., and Mufti, A. A.: The state of the art in analysis of cellular and voided slab bridges, *Canadian Journal of Civil Engineering*, 8(3), 1981, pp. 376 – 391.

7. Basu, A. K., and Dawson, J. M.: Orthotropic sandwich plates, supplement to *Proceedings, Institution of Civil Engineers*, 1970, pp. 87 – 115.

8. Cassell, A. C., Hobbs, R. E., and Basu, A. K.: *Properties of Voided Slabs*, unpublished report, Department of Civil Engineering, Imperial College of Science and Technology, University of London, London, 1970.

9. Csagoly, P. F., and Taylor, R. J.: A structural wood system for highway bridges, *Proceedings P-35/8, International Association for Bridge and Structural Engineering*, Vienna, Austria, 1980.

10. Elliott, G.: Discussion on stiffness parameters, Highway Engineering Computer Branch, Department of the Environment, *Conference on Computerized Bridge Design*, Bristol, England, 1975, pp. 4/48 – 4/53.

11. Goodman, J. R., and Bodig, J.: Orthotropic elastic properties of wood, *Journal of the Structural Division, ASCE*, 96(ST11), 1970, pp. 2301 – 2319.

12. Hambly, E. C.: *Bridge Deck Behaviour*, Chapman and Hall, London, 1976.

13. Jackson, N.: The torsional rigidity of concrete bridge decks, *Concrete*, 2(11), November 1968, pp. 468 – 474.

14. Jaeger, L. G., and Bakht, B.: The grillage analogy in bridge analysis, *Canadian Journal of Civil Engineering*, 9(2), 1982, pp. 224 – 235.

15. Kennedy, J. B., and Bali, S. K.: Rigidities of waffle-type slab structures, *Canadian Journal of Civil Engineering*, 9(1), 1979, pp. 65 – 74.

16. Ministry of Transportation and Communications: *Ontario Highway Bridge Design Code (OHBDC)*, 2d ed., Downsview, Ontario, Canada, 1983.

17. Taylor, R. J., Batchelor, B. de V., and Van Dalen, K.: Prestressed wood bridges, *International Conference on Short and Medium Span Bridges*, Toronto, Ontario, Canada, August 1982.

3

BRIDGE RESPONSES

3.1 Introduction

The term *bridge responses* refers to such quantities as the following:

1. Longitudinal bending moment
2. Longitudinal shear
3. Longitudinal twisting moment
4. Transverse bending moment
5. Transverse shear
6. Transverse twisting moment

It is noted that these quantities are often given per unit width or per unit length. For example, in a solid slab bridge the longitudinal bending moment will usually be given for a l-m- or a 1-ft-wide strip of the slab, running in the longitudinal direction. Sometimes, when the cross section of the bridge has a repetitive pattern, the bridge response is more conveniently given for "one repetition" of the pattern. For example, in the case of a slab-on-girder bridge it is often convenient to consider one girder together with that part of the slab which "belongs with" that girder (i.e., the portion of slab up to the midpoint between the girder and its neighbors on either side), and to consider longitudinal bending moments and the like in terms of this portion of the total bridge.

For load factor design, and for the ultimate limit state in limit state design, the designer works with the resistances of components, e.g., the shear capacity of a beam. By contrast, for working stress design and for the serviceability limit state of fatigue, the designer is concerned with the determination of stresses. There are some "refined" methods of analysis, such as the finite element method, which are capable of giving stresses directly; such methods may or may not include the further step of integrating the stresses so as to give responses such as bending moments and shear forces. It should be noted that the simplified methods of analysis deal directly with the approximate distribution of the responses throughout the bridge, and do not proceed by way of determination of stresses and their subsequent integration.

The responses which need to be determined for nearly all bridges are longitudinal bending moment, longitudinal shear, and transverse bending moment (including, if the deck slab has an overhang, cantilever transverse moments). To this list should be added transverse shear for bridges of the multibeam and similar types, in which the distribution of loads transversely is achieved by shear forces between adjacent longitudinally spanning elements rather than by transverse flexure. The need to calculate twisting moments arises when such factors as heavy skew are present.

For the approximate methods of analysis to succeed it is necessary that the responses concerned be capable of being calculated separately, i.e., that they should not interact with one another in a "coupled" way. Fortunately, for many types of bridges this is very nearly the case.

An advantage of using the simplified methods is that they give the designer a direct insight into the physical behavior of the structure. The physical significance of the various responses, and methods for obtaining stresses from them, are discussed in the following sections.

3.2 Longitudinal Moments

Bending moments which cause flexure in a longitudinal vertical plane are here referred to as *longitudinal moments*. It is noted that the longitudinal direction corresponds to the direction of traffic flow on a bridge. Thus, in a bridge with a rectangular planform, the longitudinal direction is parallel to the free edges of the bridge.

In an idealized orthotropic plate, bending and twisting moments are referred to per unit width or per unit length. Longitudinal bending

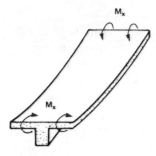

Figure 3.1 A portion of a T-beam bridge under the action of longitudinal moment M_x.

moments per unit width are designated as M_x. The physical nature of longitudinal bending moments can be understood with reference to Fig. 3.1, which shows a longitudinal strip of a T-beam bridge under the action of M_x.

Many of the simplified methods given in this book are developed by idealizing the real bridge as an orthotropic plate, which is a plate of constant thickness but having different flexural and torsional rigidities in two mutually perpendicular directions. A familiar example of an orthotropic plate is a rectangle of plywood; the grains of the top and bottom laminates run in one direction, and those of the middle laminates run in the other perpendicular direction (see Fig. 3.2). It can readily be appreciated that because of differences in the moduli of elasticity of wood in the two directions (see Sec. 2.2), the flexural rigidities of the board will be different in the two directions even though the plate thickness is uniform.

Idealization

It is important to appreciate the two-stage process involved in simplified bridge analysis: first, idealizing a structure into an orthotropic plate for obtaining responses such as longitudinal moments; and second, for obtaining stresses, reverting to another mathematical model which resembles the real structure more closely than the orthotropic plate. This

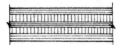

Figure 3.2 Plywood, a physical example of an orthotropic plate.

process can be explained with reference to Fig. 3.3, which illustrates longitudinal moments due to a central concentrated load at a transverse section of a multispine bridge.

Figure 3.3a shows the longitudinal flange stresses in the three-spine bridge as given by a finite strip analysis (e.g., see Ref. 7). Since the idealization employed in this method is three-dimensional and very close to the real structure, the stresses shown in Fig. 3.3a may be taken to be closely representative of the actual stresses in the structure. It can be seen that the stresses across both the top and bottom flanges are noticeably nonuniform. Analysis by idealizing the structure as an orthotropic plate gives the transverse distribution of M_x shown in Fig. 3.3b. To obtain the total moments taken by each spine one must integrate M_x over the relevant width S of the moment diagram, S being the spine spacing. Thus

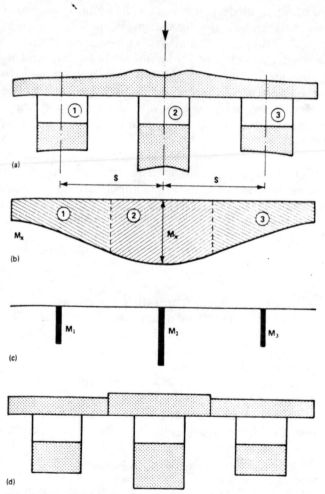

Figure 3.3 Longitudinal moments and flange stresses in a multispine bridge: (a) longitudinal flange stresses by finite strip method; (b) M_x by orthotropic plate method; (c) longitudinal moment/spine obtained from M_x; (d) longitudinal flange stresses obtained from moment/spine.

the shaded area under the moment curve identified by the number 2 in Fig. 3.3b is equal to the total longitudinal moment taken by girder 2. In a simplified method the maximum value of M_x is obtained by averaging the moment intensity over a predetermined width, and then multiplying the averaged maximum M_x by the girder, or spine, spacing. The spine moments thus obtained, M_1, M_2, and M_3, are shown in Fig. 3.3c. Once the total moment taken by a spine has been obtained, the average longitudinal flange stress σ is obtained by the familiar expression

$$\sigma = \frac{M_n z}{I} \tag{3.1}$$

where M_n is the total spine, or girder, moment; I is the second moment of area about the neutral axis; and z is the distance of the middle surface of the flange under consideration from the neutral axis of the section.

Longitudinal flange stresses obtained from moments M_1, M_2, and M_3 are shown in Fig. 3.3d. Clearly, the orthotropic plate analysis is unable to pick up the local nonuniformity of the stresses. In the case of eccentric loading, even the moments taken by the two webs and associated flange widths of the same spine may be substantially different from one another, thus making the orthotropic plate analysis a little further removed from reality.

The departure of orthotropic plate theory from reality becomes less significant when looked at in terms of total load effects due to live load and dead load. For example, if live and dead loads are about the same, then the percentage of error due to live-load analysis is roughly halved in terms of total load effects.

Effective Width

The local increase of longitudinal flexural stress in the portion of a flange closest to a web is caused by the well-known shear lag phenomenon. It is recalled that this phenomenon refers to the diminution of longitudinal stresses in flanges of such sections as T beams as one moves away from the junction of the flange and web. As was noted in Sec. 2.2, the shear lag effects do not have a substantial effect on the flexural rigidity of the T section. Hence, the share of total moment taken by a given section remains almost independent of the shear lag phenomenon. The local increase in flange stresses is, however, taken into account during the calculation of stresses from moments, by using a reduced flange width. This reduced width is often referred to as the *effective flange width*.

By analyzing a large number of slab-on-girder and multispine bridges by the finite strip method, it has been confirmed (see Ref. 6) that the effective width depends directly on the ratio of L/B, where the notation is as shown in Fig. 3.4. It has also been confirmed that the effective flange width is largest when the beams are subjected to uniformly distributed loads and smallest when the beams are subjected to central concentrated loads. These two upper and lower bounds of effective widths are shown in Fig. 3.4, in which the ratio B_e/B is plotted against L/B. Figure 3.4 shows the curve for effective widths due to AASHTO [1] and Ontario [9] live loadings as recommended in Ref. 6. The Ontario code has simplified the calculation of effective width by combining the effective widths for dead and live loads. The Ontario expression, which was developed from analysis results given in Ref. 6 and by assuming typical ratios of dead and live loads in various bridges, is as follows:

$$\frac{B_e}{B} = 1 - \left(1 - \frac{L}{15B}\right)^3 \quad \not> 1.0 \tag{3.2}$$

The notation is the same as illustrated in Fig. 3.4, which also shows the curve plotted from Eq. (3.2).

For T-beam concrete bridges, the AASHTO specifications [1] require that no shear lag effects be taken into account, i.e., that B_e shall be

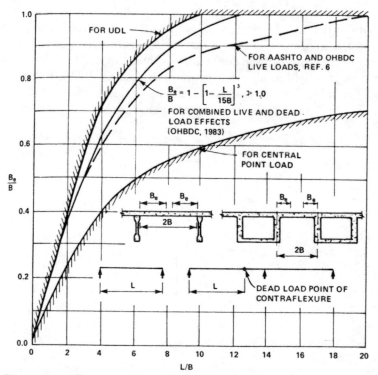

Figure 3.4 Effective widths. *(Adapted from Ref. 6.)*

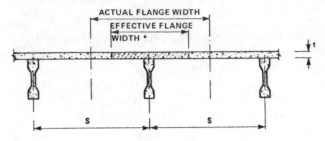

*LEAST OF ACTUAL FLANGE WIDTH, L/4 AND 12t (AASHTO, 1977)

Figure 3.5 Effective flange widths according to AASHTO specifications.

assumed equal to B. For other structures the effective width, as shown in Fig. 3.5, should be the smallest of the following:

1. One-fourth of the span
2. Center-to-center spacing of girders
3. Twelve times the deck slab thickness

Effect of Transverse Load Position

The maximum intensity of longitudinal moments in an orthotropic plate representing a bridge increases with increasing eccentricity of the load with respect to the longitudinal centerline of the bridge. This is demonstrated in Fig. 3.6, which shows a transverse plot of the distribution coefficients for M_x under a concentrated load in a slab bridge. (A distribution coefficient is the ratio of the local value of M_x to the average

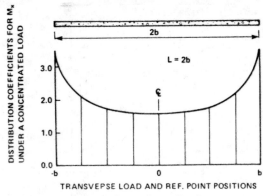

Figure 3.6 Effect of transverse load position on longitudinal moment under load.

intensity of longitudinal moment $M_{x,av}$ on the cross section.) It can be seen from the figure that moving a load from the middle to the free edge of the bridge can cause the maximum intensity of longitudinal moments to increase by up to 125 percent. This figure is, of course, true only for single loads of very small contact area. For multiple loads and loads of larger contact area, as on real bridges, the increase would be smaller. However, the fact remains that maximum intensities of longitudinal moments increase with load or vehicle eccentricity, and that the increase is more pronounced when the load is on the outer quarter width of the bridge.

3.3 Longitudinal Shear

The term *longitudinal shear* refers to the vertical shear in a longitudinal strip of the bridge. In simply supported bridges longitudinal shear is that bridge response which at the supports can be regarded as the support reaction. This response, the intensity per unit width of which is designated by V_x, is illustrated in Fig. 3.7. The figure shows the action of V_x at a transverse section in a longitudinal strip of a T-beam bridge.

The transverse distribution of longitudinal shears is markedly different from that of longitudinal moments, the latter being more benign, i.e., less "peaky." Differences in the distribution patterns of the two responses can be observed in Fig. 3.8, which shows the results of a grillage analysis on a slab-on-girder bridge. It can be seen that the distribution of V_x, which corresponds to a maximum distribution coefficient of 3.11, is reasonably independent of the longitudinal position of the section under investigation. Similarly, the transverse distribution pattern of longitudinal moments is substantially independent of the longitudinal position

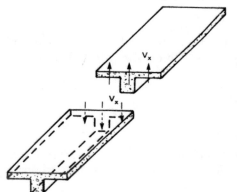

Figure 3.7 Portions of a T-beam bridge illustrating longitudinal shear V_x.

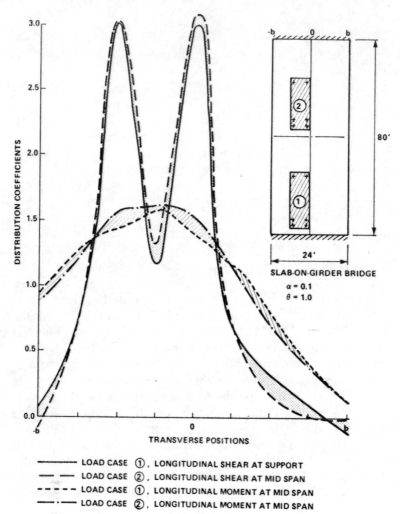

Figure 3.8 Transverse distributions of longitudinal shears and moments.

of reference points. The maximum value of the distribution coefficient for bending moment, however, is only 1.6, i.e., about half that for V_x.

3.4 Transverse Moments

Bending moments in the direction perpendicular to the flow of traffic, but excluding moments in cantilever deck slab overhangs, are referred to in this book as *transverse moments*. A transverse slice of a slab-on-girder bridge, under the action of a transverse moment of intensity M_y, is shown in Fig. 3.9.

There is a great deal of uncertainty about how to make an accurate determination of this response. For the same bridge and the same loads,

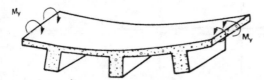

Figure 3.9 A portion of a T-beam bridge under the action of transverse moment M_y.

two different refined methods of analysis are likely to predict markedly different values for M_y. Yet the same two methods are likely to give almost identical longitudinal moments and shears. The determination of live-load transverse moments becomes even more uncertain in the case of bridges with concrete deck slabs on flexurally stiff members, such as girders or webs.

Local and Global Moments

For the purpose of determining live-load transverse moments in the deck slabs of slab-on-girder bridges, or in bridges with webs (for example, cellular and box girder bridges) it is customary to divide the total response into two components, namely, local moments M_{yL}, and global moments M_{yg}. The former response is obtained by assuming that the girders or webs do not deflect. This assumption reduces the problem to that of a plate supported on a number of parallel and unyielding supports. The global moments are obtained in one of two ways. In the first of these the bridge superstructure is regarded as a grillage, with the grillage longitudinal beams being coincident in position with the girders or webs of the bridge. The loads which are situated between girders are replaced by a statically equivalent set of loads placed on adjacent girders. This concept is illustrated in Fig. 3.10.

In the second approach to global moments the bridge superstructure is regarded as an orthotropic plate; i.e., the longitudinal stiffnesses are treated as uniformly distributed across the width of the bridge. The loads remain in the same positions. This concept is illustrated in Fig. 3.11b, where the uniform distribution of stiffnesses is approximated by a relatively large number of closely spaced elastic supports.

Neither of these two approaches is fully defensible mathematically. Indeed, as noted above, they lead to significantly different results, both

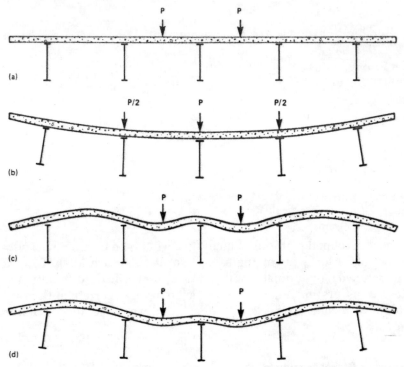

Figure 3.10 Local and global transverse moments: (*a*) undeformed cross section; (*b*) deformations due to global moments (loads assumed to be acting only on girders); (*c*) deformations due to local moments (girders assumed to be nondeflecting); (*d*) assumed actual deformations (due to local and global moments).

from one another and from the moments which result from exact analysis. The first approach usually leads to overestimation of the maximum negative transverse moments and underestimation of the maximum positive ones. The second approach usually overestimates both.

To demonstrate that the second approach is not compatible with the actual behavior of the structure, an analogy is drawn between a transverse slice of a slab-on-girder bridge and a beam on discrete elastic supports. Such a beam on four spring supports, representing a transverse slice of a bridge with four girders, is shown under a central point load in Fig. 3.11*a*. Beam moments for this condition are also shown in Fig. 3.11*a*. Global moments in the beam according to the second approach are obtained by increasing the number of spring supports to seven, thereby providing a support condition reasonably close to plate action, and evenly distributing the total stiffness of the original four springs among the seven new supports. Beam moments corresponding to this condition are shown in Fig. 3.11*b*. Those corresponding to the local moment condition, in which the springs are replaced by unyielding supports, are shown in Fig. 3.11*c*.

It can be seen in Fig. 3.11*d* that the sum of local and global moments is significantly higher than the actual moments. This confirms that the practice of splitting transverse moments into global and local moments in this manner is an approximation which leads to conservative results.

Arching Effect

Extensive laboratory and field testing of deck slabs of slab-on-girder bridges in Ontario [2, 4] and in New York [5] has conclusively shown that these slabs possess far more strength than could be possible if transverse moments as predicted by conventional analyses did really exist. The very significant enhancement of the deck slab behavior is reportedly caused

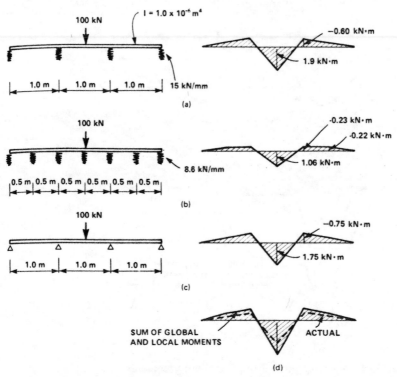

Figure 3.11 Beam on spring supports: (*a*) Actual condition. (*b*) Condition corresponding to global moments. (*c*) Condition corresponding to local moments. (*d*) Comparison of actual moments with those obtained by adding global and local moments.

by an arching action which is difficult to predict with any great degree of accuracy even by means of very refined analytical tools.

This very large discrepancy between transverse moments predicted by the usual, albeit refined, analyses and the actual transverse moments in the deck slab of a slab-on-girder bridge is illustrated in Fig. 3.12. The figure shows measured deflections of a slab-on-girder bridge under a concentrated load simulating a double-tire footprint. The results of the test are reported in Ref. 8. From measured deflections, transverse curvatures and thence transverse moments were calculated without accounting for the curvature in the longitudinal direction. Thus, these moments may be regarded as those corresponding to a Poisson's ratio of zero. The

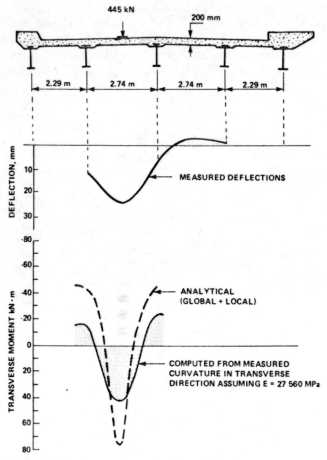

Figure 3.12 Comparison of measured transverse moments with analytical local moments.

analytical moments shown in the figure are the sum of local and global moments, the former being obtained for a Poisson's ratio of zero.

It is recognized that small errors in measurements of deflections can result in large errors in curvature, and also that moments computed from measured curvatures depend upon an estimated value of Young's modulus. However, these errors alone cannot account for the large difference between measured and calculated moments. Thus, it is important to realize that in bridges with concrete deck slabs on girders the transverse moments obtained even by refined methods are, at best, very conservative estimates of the real response.

Effect of Transverse Load Position

For a single concentrated load on a given transverse line, the maximum transverse moments are induced when the load is midway between the free edges of the bridge. Accordingly, maximum sagging transverse moments (causing tension in bottom fibers) due to a vehicle are induced when the vehicle is on the bridge centerline. The vehicle position causing the maximum local sagging moment may not correspond to the vehicle position which produces the maximum global sagging moment. Therefore, the practice of adding the two maximum effects will usually be somewhat conservative.

Hogging global transverse moments occur only in very wide bridges when two vehicles are symmetrically placed as close as possible to the free edges. The global hogging, or negative, moments induced in this manner occur in the unloaded middle portion of the bridge. Therefore, negative local moments which occur over girders in the vicinity of loads cannot be added to the global negative moments. The latter are usually so small that the steel provided in the deck slab for shrinkage and temperature effects is sufficient to accommodate them.

3.5   Transverse Cantilever Moments

The problem of overhanging concrete deck slabs of slab-on-girder and box girder bridges can be simplified by treating the overhanging slab as a cantilever slab which is fixed against both rotation and vertical deflection at its root, i.e., over the centerline of the outer girder or web. Transverse cantilever moments in a deck slab overhang due to a concentrated load are shown in Fig. 3.13.

It is noted that the girder flexibility has a beneficial effect on the longitudinal distribution of the cantilever moments due to a single concentrated load. However, as shown in Ref. 10, the reduction of peak cantilever moments due to support flexibility is small in most cases, being of the order of 5 percent.

Effect of Load Position

When a concentrated load is at a distance of at least $2a$ from a transverse free edge of the deck slab overhang, where a is the length of the overhang in the transverse direction, the maximum cantilever moment is induced above the girder at a point which is transversely on the same line as the load (see Fig. 3.13). However, when the load is in the vicinity of transverse free edge, the maximum cantilever moment occurs over the girder at the free edge, and the intensity of this moment is significantly larger than the corresponding maximum moment due to a load away from the transverse free edge.

As pointed out in Ref. 3, sagging moments, both in the longitudinal and transverse directions, do occur under concentrated load. However, these moments are localized to the vicinity of the load, and are small enough to be neglected.

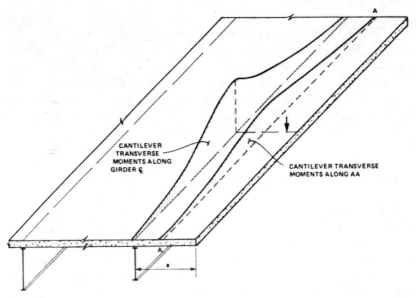

Figure 3.13   Cantilever transverse moments in the deck slab of a slab-on-girder bridge.

3.6 Transverse Shear

As discussed in Sec. 2.3, load transfer in multibeam bridges, from one beam to an adjacent one, takes place through shear keys which are placed between the beams. The shear force in these keys is here referred

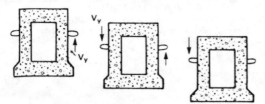

Figure 3.14 Transverse sections of beams in a multibeam bridge.

to as *transverse shear*. This response is diagrammatically shown in Fig. 3.14, and its intensity per unit length is designated as V_y.

One aspect of this response worth noting is its highly localized nature. The intensity of transverse shear is a maximum in the immediate vicinity of the load. In the vicinity of a concentrated load, the peak intensity of V_y is usually concentrated over a length which is about one-twelfth of the span.

The intensity of transverse shear, which is maximum when the load is closest to the longitudinal free edge of the bridge, is not appreciably affected by the longitudinal position of the load. Therefore, all shear keys along a longitudinal joint in a multibeam bridge should be designed to have the same capacity. It is, however, logical, albeit impractical, to have weaker shear keys in joints toward the middle of the bridge than in those nearer the free edges.

References

1. American Association of State Highway and Transportation Officials (AASHTO): *Standard Specifications for Highway Bridges*, Washington, D.C., 1977.
2. Bakht, B., and Csagoly, P. F.: *Bridge Testing*, Structural Research Report SRR-79-10, Ministry of Transportation and Communications, Downsview, Ontario, Canada, 1979.
3. Bakht, B., and Holland, D. A.: A manual method for the elastic analysis of cantilever slabs of linearly varying thickness, *Canadian Journal of Civil Engineering*, 3(4), 1976, pp. 523–530.
4. Batchelor, B. de V., Hewitt, B. E., Csagoly, P. F., and Holowka, M.: *An Investigation of the Ultimate Strength of Deck Slabs of Composite Steel/Concrete Bridges*, TRR 664, Transportation Research Board, Washington, D.C., 1978.

5. Beal, D. B.: Load capacity of concrete bridge decks, *Journal of Structural Division, ASCE*, O8(ST4), 1982, pp. 815–832.
6. Cheung, M. S., and Chan, M. Y. T.: Finite-strip evaluation of effective flange width of bridge girders, *Canadian Journal of Civil Engineering*, 5(2), 1978, pp. 174–185.
7. Cheung, M. S., Cheung, Y. K., and Ghali, A.: Analysis of slabs and girder bridges by the finite strip method, *Building Science*, 5, 1970, p. 95.
8. Csagoly, P. F.: *Design of Thin Concrete Deck Slabs by the Ontario Bridge Design Code*, Structural Research Report SRR-79-11, Ministry of Transportation and Communications, Downsview, Ontario, Canada, 1979.
9. Ministry of Transportations and Communications: *Ontario Highway Bridge Design Code (OHBDC)*, 2d ed., Downsview, Ontario, Canada, 1983.
10. Sawko, F., and Mills, J. H.: Design of cantilever slabs for spine beam bridges: Discussion, *International Conference on Developments in Bridge Design and Construction*, University College, Cardiff, Wales, 1971.

4

METHODS FOR LONGITUDINAL MOMENTS IN SHALLOW SUPERSTRUCTURES

4.1 Introduction

As defined earlier, the term *shallow superstructure* is used for the following bridge types:

1. Slab
2. Voided slab
3. Slab-on-girder

The simplified methods for determining longitudinal moments presented in this chapter are modeled on the AASHTO method [1], which is familiar to most bridge designers in North America. The AASHTO method for determining live-load moments in a slab-on-girder bridge requires that a girder plus its associated portion of the slab be subjected to a load comprising one line of wheels of the design vehicle, with the wheel loads multiplied by a fraction S/D, where S is the girder spacing and D has a prespecified value depending upon the type of bridge. (This method is diagrammatically shown in Fig. 4.1.) For example, for a bridge with steel girders and a concrete deck slab and having two or more lanes, D is specified to be 1.676 m (5.5 ft). In the case of slab bridges, the same

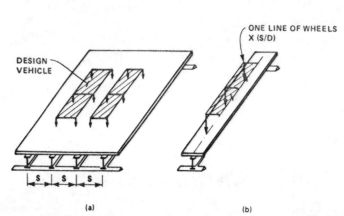

Figure 4.1 Illustration of the D-value method: (a) actual loading; (b) loading used for calculating girder moment.

procedure is followed, except that S becomes a unit width and the moments obtained correspond to a unit width of the slab.

D-Type Methods

A D-type method, which is usually developed by idealizing a bridge as an orthotropic plate, is based upon the premise that the distribution pattern of intensity of longitudinal moments across a transverse cross section is reasonably independent of the longitudinal position of the load and the transverse section considered. This premise is not new and, as discussed in Sec. 1-10, has been the basis of other well-established simplified methods (e.g., Refs. 5 and 8). The D-type method, however, is the simplest to use, in the opinion of many designers.

The physical significance of the above-mentioned premise can be examined with the help of Fig. 4.2, which shows the transverse distribution of longitudinal moments at two different transverse sections in a slab bridge subjected to a vehicle load. Let the total longitudinal moments at sections 1-1 and 2-2 be M_1 and M_2, respectively, and let the intensities of longitudinal moments at these two sections along the same longitudinal line be designated M_{x1} and M_{x2}, respectively. The premise that the distribution pattern of longitudinal moments is fairly independent of the longitudinal position of the reference section implies that the ratio M_{x1}/M_1 is nearly equal to M_{x2}/M_2. For bridges that are free of skew, this condition is met sufficiently closely for design purposes.

The concept of the factor D can be explained with reference to Fig. 4.3, which shows the transverse distribution of M_x at a cross section due to a vehicle on a slab-on-girder bridge idealized as an orthotropic plate.

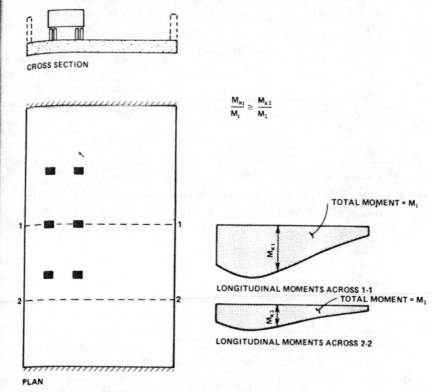

$$\frac{M_{x1}}{M_1} \cong \frac{M_{x1}}{M_2}$$

Figure 4.2 Distribution patterns for longitudinal moments.

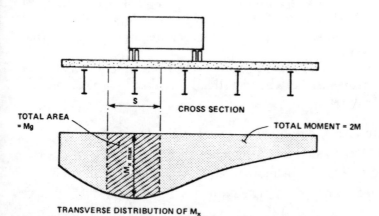

Figure 4.3 Longitudinal moment distribution across a transverse section.

The total live-load moment M_g that the second girder from the left (along with its associated portion of the deck) should be required to sustain is equal to the area under the curve shown hatched in the figure. If the intensity of the maximum longitudinal moment is $M_{x,max}$, and the girder spacing is equal to S, then this area is approximately equal to $S \times M_{x,max}$. Thus:

$$M_g = SM_{x,max} \tag{4.1}$$

Now assume that the unknown quantity $M_{x,max}$ can be obtained by a factor D which is given by

$$D = \frac{M}{M_{x,max}} \tag{4.2}$$

where M is one-half the total moment due to one vehicle, or, in other words, the moment due to one line of wheels at the section under consideration. Substituting the value of $M_{x,max}$ from Eq. (4.1) into Eq. (4.2), we get:

$$M_g = \frac{S}{D} M \tag{4.3}$$

Thus, if the value of D is known for a bridge, the maximum live-load moment in a width S of the bridge can be obtained as a fraction S/D of the beam bending moment due to one line of wheels. The spacing S represents the "pattern" of the cross section, if one exists. For a slab bridge S is taken to be unit width.

The degree of approximation inherent in Eq. (4.1) can be significantly reduced if $M_{x,max}$, which is used for deriving values of D through Eq. (4.2), is obtained by averaging M_x over a finite width. For the methods given in this chapter, values of $M_{x,max}$, and therefore of D, were obtained by averaging M_x in the vicinity of the peak moment over a width of approximately 2 m (6.56 ft). Wheel loads were given contact areas of 0.81×0.46 m (2.66×1.51 ft), with the larger dimension running in the transverse direction.

The principal factors which can affect the transverse distribution of live load in a bridge that can be realistically idealized as an orthotropic plate are as follows:

1. Longitudinal and transverse flexural rigidities of the bridge
2. Longitudinal and transverse torsional rigidities of the bridge
3. Aspect ratio of the planform, i.e., the ratio of span to width
4. Type of design loading (e.g., uniformly distributed, knife-edge, or truck load)

5. Width of the load with respect to the bridge width

6. Number of design loadings in the transverse direction

7. Vehicle-edge distance, i.e., the transverse distance between a longitudinal free edge of the bridge and the nearest load

The AASHTO method [1] groups all the above factors into a single D value which is constant for a given type of bridge. This oversimplification can give erroneous results for some cases. The methods given in this book also utilize the D-value concept; however, the values of D are based on explicit considerations of the factors listed above, and can change in bridges of the same type.

Limitations on Bridge Geometry

The simplified methods given in this chapter are developed by idealizing bridges as orthotropic plates. In order that a bridge can be satisfactorily idealized as an orthotropic plate, it must satisfy the following conditions reasonably closely:

1. The width is constant.

2. The support conditions are closely equivalent to line supports.

3. The skew angle does not exceed 20°.

4. In the case of curved-in-plan bridges, L^2/bR is smaller than 1.0, where the notation is as shown in Fig. 4.4d.

5. A solid or voided slab bridge is of uniform depth across the cross section.

6. The total flexural rigidity of the transverse cross section remains the same for at least the central 50 percent of each span.

7. For slab-on-girder bridges there are at least four girders, equally spaced and of equal flexural rigidity.

8. For cellular bridges there are at least three equal cells.

9. For slab-on-girder bridges having an overhanging deck slab, the overhang does not exceed 60 percent of the spacing between girders and also is not more than 1.8 m (5.96 ft); further, the outermost extremity of a design lane is not more than 1.0 m (3.28 ft) outside the centerline of the outermost girder.

Figure 4.4 illustrates some of the more common ways in which the above limitations may be violated. These limitations are, of course, intended only as guides. Few, if any, real-life bridges can precisely satisfy these conditions. Engineering judgment is expected to be exercised to determine whether a bridge meets these conditions sufficiently closely for idealization as an orthotropic plate to be reasonable, and hence for the simplified methods to be applicable. To assist in the exercise of engineering judgment, a discussion of some of the conditions follows.

The requirement for the support to be closely equivalent to a line support is applicable only when the response being investigated is in the vicinity of the support, e.g., longitudinal moments over interior supports of a multispan bridge, or longitudinal shears near a support. The support condition of a slab-on-girder bridge, in which all longitudinal girders are supported, can be regarded as a satisfactory representation of a line support. Discrete supports in slab and voided slab bridges can be regarded as line supports if they are spaced not more than 2.0 m (5.96 ft) apart. The stipulation of 2.0 m (5.96 ft) is based on the fact that the orthotropic plate longitudinal moments and shears are averaged over 2.0-m (5.96-ft) widths when obtaining the governing D values, as discussed later.

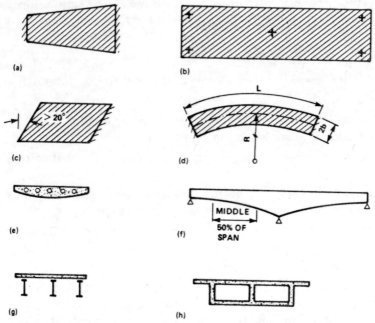

Figure 4.4 Bridges that cannot be analyzed by the simplified methods: (a) bridge with varying width; (b) bridge with few isolated supports; (c) skew bridge with skew angle greater than 20°; (d) curved-in-plan bridge with $L^2/bR > 1.0$; (e) bridge with nonuniform cross section; (f) bridge with nonuniform section in the middle 50 percent of span (except as noted in Chap. 8); (g) slab-on-girder bridge with fewer than four girders; (h) cellular bridge with fewer than three cells.

When a skew bridge is analyzed, its skew span should be used in the analysis as if it were the right span. The limitation of 20° on the skew angle is imposed to ensure that twisting moments, which are customarily not calculated directly in simplified analyses, are not so large as to be troublesome. The limitation on the skew angle can be eased for slab-on-girder bridges, up to about 30°.

The analogy between a slab-on-girder bridge and an orthotropic plate becomes dubious when there are fewer than four girders — hence, the limitation on the number of girders, and similarly on the number of cells in cellular structures. The analysis of a slab-on-girder bridge having only two girders can be carried out in the manner discussed in Chap. 12.

When a load is outside of the outermost girder in a slab-on-girder bridge, the orthotropic idealization of the structure becomes suspect and may give results which are on the unsafe side. The limitation of the design lane being within 1.0 m (3.28 ft) of the outermost girder ensures that the method based on orthotropic plate idealization is not used for such cases.

The above limitations on bridge geometry are illustrated in Fig. 4.4. It is noted that the restriction on variation of longitudinal flexural rigidity (Fig. 4.4f) can be relaxed, as discussed in Chap. 8.

Developmental Background

The methodology for the simplified methods presented in this chapter was developed for the *Ontario Highway Bridge Design Code*, first and second editions [6, 7]. The development of this methodology is described in Ref. 4. Unlike other bridge design codes, the Ontario code has an entire section devoted to specifying permissible methods of bridge analysis. One of the objectives of the analysis section of the code was to develop a simplified method which would not be significantly different in its general nature from the AASHTO method, but yet would be able to give more accurate results. The method now specified in the Ontario code does satisfy these requirements.

Earlier development of the method, which is based on the orthotropic plate theory [3], is discussed in Ref. 2. The method as given in the first edition of the *Ontario Highway Bridge Design Code* has undergone substantial changes which are, as yet, unreported in the literature. The revised method is specified in the second edition [7].

External and Internal Portions

The methods for Ontario and AASHTO loadings, given in Secs. 4.2 and 4.3, differentiate between external and internal portions of the cross

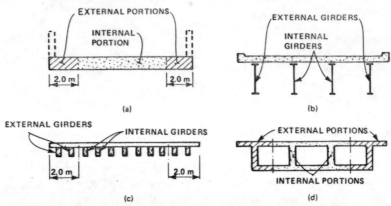

Figure 4.5　Demarcation of external and internal portions on bridge cross section. (*a*) Slab bridge. (*b*) Slab-on-girder bridge. (*c*) Sawn-timber stringer bridge. (*d*) Cellular structure.

section of a bridge, and provide different D values for the two. External portions in slab and voided slab bridges are the outermost 2-m (6.56-ft) strips of the bridge, as shown in Fig. 4.5a. The internal portion is, of course, that part of the transverse cross section which is contained between the two external portions.

In slab-on-girder bridges having girders spaced at more than about 2.0 m (6.56 ft), the external portions refer to the outer girders, and internal portions to the inner ones. However, when the girder spacing is much smaller than 2.0 m (6.56 ft), as in wood bridges with sawn stringers, then all girders within the outer 2-m (6.56-ft) strips should be regarded as external. The adoption of a 2-m-wide (6.56-ft-wide) strip as defining the external portion of a cross section is in line with the analytical procedure used in deriving the simplified method. In this procedure the longitudinal bending moments in the outermost 2 m (6.56 ft) of the cross section of the corresponding orthotropic plate are averaged in order to arrive at the D value applicable to this part of the cross section.

The division of the cross section into internal and external portions is shown in Fig. 4.5, for four types of bridges.

4.2　Method for Ontario Loading

The Ontario highway bridge design code permits the use of the method given here, for bridges which conform to the limitations given in Sec. 4.1 and illustrated in Fig. 4.2.

Ontario Loading

The Ontario design loading is of two types:

1. A five-axle truck with two lines of wheels 1.8 m apart

2. A uniformly distributed load and the same five-axle truck as in type 1 but with the axle loads at 70 percent of the former level

The former loading is designated as the *design truck loading* and the latter as the *design lane loading*. Each lane of a bridge is required to be

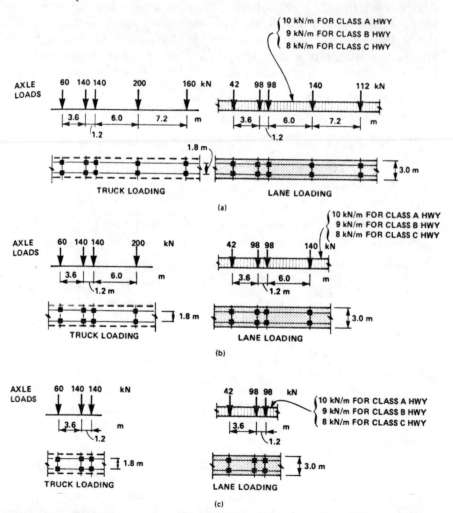

Figure 4.6 Ontario design and evaluation loads: (*a*) design loading and level 3 evaluation loading; (*b*) level 2 evaluation loading; (*c*) level 1 evaluation loading.

loaded by either the truck load or the lane load, according to which produces greater load effects. Details of the design loadings are shown in Fig. 4.6, which also shows the evaluation loads used for determining the load-carrying capacities of existing bridges. There are three levels of evaluation loads. The third-level loadings are the same as the design loadings (Fig. 4.6a). Second-level loadings are the same as level 3 loadings, except that one axle is dropped (Fig. 4.6b). Dropping yet another axle produces the level 1 loadings (Fig. 4.6c).

The lane loadings are really meant for longer spans, where maximum load effects are caused by more than one vehicle in a lane. The uniformly distributed lane load obviates the necessity for having more than one design truck in a lane, regardless of the span length and the number of spans. As shown in Fig. 4.6, the uniformly distributed load varies with the highway class. For the purposes of bridge design, class A highways are those with a minimum average daily truck count of 1000; class B highways are those with an average daily truck count between 250 and

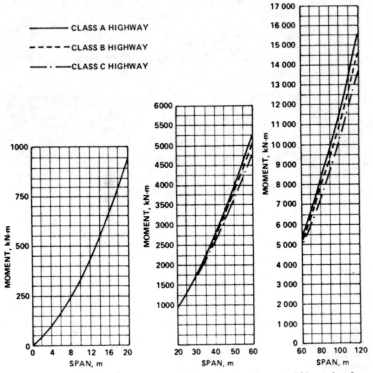

Figure 4.7 Moments due to one line of wheels or half-lane loading, OHBDC.

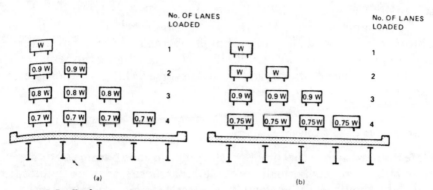

Figure 4.8 Reduction factors for multilane loading: (a) Ontario; (b) AASHTO.

1000; highways with a daily truck count of less than 250 are designated as class C.

Maximum simple span bending moments due to one line of wheels of the truck loading, or one-half of the lane loading, are plotted in Fig. 4.7 for various span lengths. It should be noted that the class of the highway has an effect on bending moments only when the span exceeds about 26 m (85 ft).

When two or more lanes of a bridge are loaded, the design loading is reduced by applying a reduction factor. When two lanes are loaded, the reduction factor is 0.9; with three lanes loaded, it is 0.8; and with four lanes loaded it is 0.7. See Fig. 4.8a.

Design Lanes

The Ontario code requires that the number of design lanes in a bridge be obtained by dividing the total curb-to-curb width into lanes of equal width according to the criteria given in Table 4.1. The design lanes

TABLE 4.1 Number of Design Lanes According to the Ontario Code

Curb-to-curb transverse bridge width, m	Number of design lanes
<6.0	1
6.0–10.0	2
10.0–13.5	3
13.5–17.0	4
17.0–20.5	5
20.5–24.0	6
24.0–27.5	7
>27.5	8

obtained in this manner are required to be used for the ultimate limit state (ULS) and serviceability limit state type II (SLS II).

In recognition of the fact that the actual number of lanes may be different from the number of design lanes, the code specifies that, in a consideration of the serviceability limit state of fatigue, the actual number of lanes on the bridge, and the transverse vehicle positions corresponding to them, shall be used.

Dynamic Load Allowance

Consistent with the usual practice, the Ontario code permits static analysis to be used for moving loads, provided that these loads are converted to equivalent static loads. The equivalent static loads are obtained by using a factor called the *dynamic load allowance* (DLA), which is frequently also referred to as the *impact factor*. The equivalent static load is equal to the actual load multiplied by (1 + DLA).

For deck slab elements which are governed by a single wheel, single axle, or the dual axle of the design vehicle, the DLA is 0.4. When the loading is other than a single wheel or a single axle, the DLA is 0.30 for all transverse members, all simple spans not more than 22 m (72.7 ft) in length, and all superstructures composed of continuous spans with no span greater than 22 m (72.7 ft).

For all main longitudinal components and those not covered above, the DLA for axle loads is as given in Fig. 4.9; it depends upon the first flexural frequency of vibration of the bridge. The DLA for the uniformly distributed portion of the design lane load is specified to be 0.10. For initial design, the commentary to the code suggests the use of the following values of the DLA for loading in one lane:

For simple spans between 22.5 and 60.0 m (73.8 and 196.9 ft):   0.40

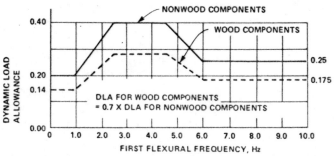

Figure 4.9 Dynamic load allowance according to the Ontario code.

For simple spans greater than 60.0 m (196.9 ft):   0.20

For continuous bridges with any span between 22.5 and
  60.0 m (73.8 and 196.9 ft):   0.40

For continuous bridges with any span greater than 60.0 m
  (196.9 ft):   0.20

Methods for determining the first flexural frequencies for simple cases are given in Appendix I. It is recommended that in the case of noncomposite slab-on-girder bridges the first flexural frequency be calculated by assuming the structure to be composite.

Simplified Method for ULS and SLS II

The Ontario code requires that for the ultimate limit state and serviceability limit state type II, as many lanes in a bridge must be loaded as will produce the maximum load effect. The method given herein was developed for one-, two-, three-, and four-lane bridges for the various loaded-lane conditions. The condition which, after the application of the relevant reduction factors (Fig. 4.8a), produces the maximum longitudinal moment is the governing one. Governing load cases in bridges with different numbers of lanes and for different values of α and θ, as defined in Sec. 2.2, are shown in Fig. 4.10. For the use of the following method, however, the designer need not know which loaded-lane condition governs the design.

The various steps in calculating live-load longitudinal moments in both external and internal portions of a bridge are given below. For design of a new bridge, all the steps should be followed; for evaluation of an existing bridge, steps 1, 2, 4, and 5 are not relevant and are omitted.

1. Obtain an initial value of D, identified as D_d, from Table 4.2 (see page 109), according to the bridge type and number of design lanes in the bridge.

2. Calculate the initial load fraction S/D_d, where S = the actual girder spacing in the case of a slab-on-girder bridge; or the spacing of webs in the case of voided slabs and cellular structures; or 1 m, or 1 ft, in the case of solid slabs, transversely prestressed laminated-wood bridges, and concrete-wood composite bridges composed of wood laminates and concrete overlays.

3. Treating the bridge as a one-dimensional beam, obtain bending moment diagrams due to one-half the truck or lane loadings shown in Fig. 4.6.

4. Multiply the moments obtained in step 3 above by $(S/D_d)(1 + DLA)$ to obtain the initial live-load moments, where DLA is the dynamic load allowance obtained as discussed earlier. [It is noted that

the modification factor for multilane loading (Fig. 4.8) is implicit in the D_d values, and should not be applied again.] In the case of lane loadings which consist of both axle loads and uniformly distributed loads, it is necessary to apply different values of the DLA to moments due to the two types of loads.

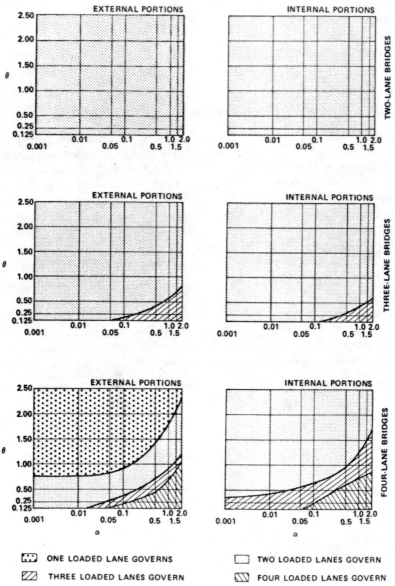

Figure 4.10 Governing load cases corresponding to Ontario loading.

5. Assume that the design live-load moments obtained in step 4 above are sustained by a width S of the bridge for both external and internal portions, and on this basis obtain initial proportions of the structure.

6. Calculate α and θ from the following expressions:

$$\alpha = \frac{D_{xy} + D_{yx} + D_1 + D_2}{2(D_x D_y)^{0.5}} \tag{4.4}$$

$$\theta = \frac{b}{L} \left(\frac{D_x}{D_y} \right)^{0.25} \tag{4.5}$$

The notation and the procedure for calculating α and θ for various bridge types are given in Sec. 2.2.

7. Calculate μ from

$$\mu = \frac{W_e - 3.3}{0.6} \qquad \mu \geqslant 1.0 \tag{4.6}$$

where W_e is the design lane width in meters.

8. Corresponding to the values of α and θ, and the number of design lanes in the bridge, obtain the values of D separately for external and internal portions and the value of C_f from relevant charts of Fig. 4.11.

9. Obtain the final values of D_d separately for external and internal portions from

$$D_d = D \left(1 + \frac{\mu C_f}{100} \right) \tag{4.7}$$

10. For each of the external and internal portions, obtain the final live-load design moments by multiplying the live-load moments due to one line of wheels or half-lane load obtained in step 3 above by $(S/D_d)(1 + \text{DLA})$, where D_d has the value appropriate to the external or internal portion. Again the note about different values of the DLA for axle and uniformly distributed loads as given in step 4 above applies.

A summary of the above procedure is given in Fig. 4.12.

Notwithstanding the above, the Ontario code requires that the D_d value for longitudinally nail-laminated wood bridges should be taken as 0.85 m (2.79 ft) for both external and internal portions. When these bridges are new, it is possible to idealize them as articulated plates and to analyze them according to the method given in Chap. 11.

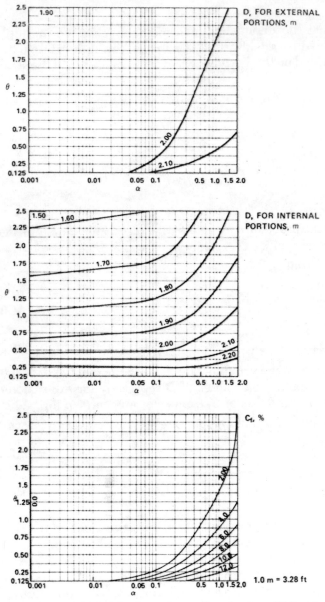

Figure 4.11a *D* and *C_f* charts for one-lane bridges, ULS and SLS I, OHBDC, and governing load case, AASHTO loading.

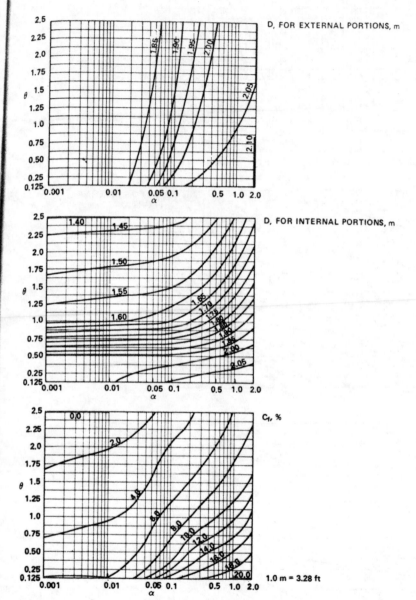

Figure 4.11*b* *D* and C_f charts for two-lane bridges, ULS and SLS II, OHBDC loading.

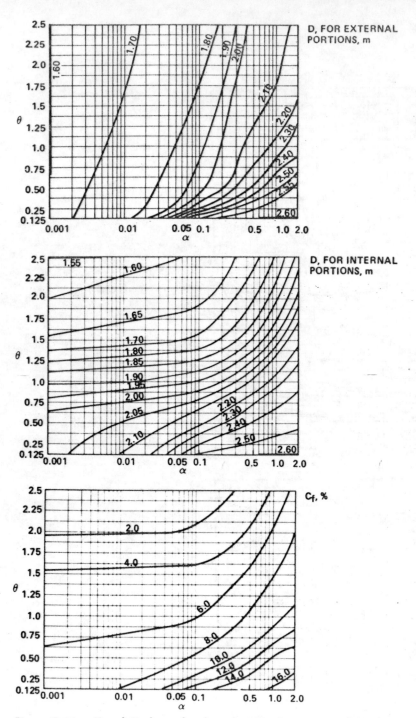

Figure 4.11c D and C_f charts for three-lane bridges, ULS and SLS II, OHBDC loading.

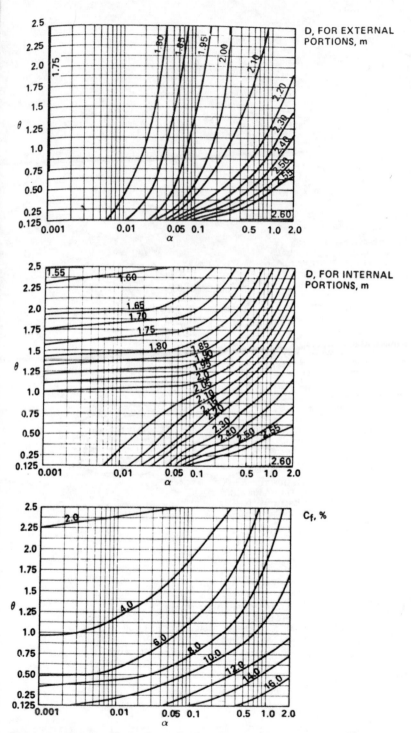

Figure 4.11d D and C_f charts for four-lane bridges, ULS and SLS II, OHBDC loading.

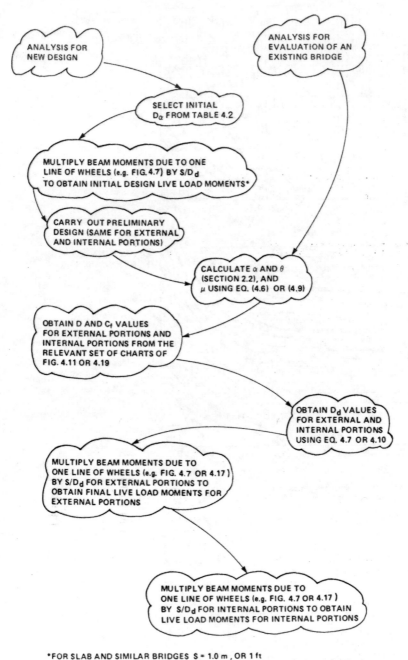

*FOR SLAB AND SIMILAR BRIDGES S = 1.0 m , OR 1 ft

Figure 4.12 Summary of procedure for obtaining live-load moments.

Simplified Method for SLS I

For serviceability limit state type I (SLS I), which includes the limit state of fatigue, the bridge is loaded in a single lane. Analysis for this limit state is different from that for ULS: the analysis is carried out after the structure is initially proportioned, and the D value is that for a single vehicle in its normal traveling transverse position. The Ontario code does not distinguish between external and internal portions of the cross section for analysis corresponding to this limit state. However, in this book, the method is given so that the designer can obtain different load effects in external and internal portions of the cross section, if desired. The designer may, of course, choose to use the moments for external portions for internal portions also. It is noted that the moments for external portions are always higher than those for internal ones.

The method now presented corresponds to the design vehicle being in the extreme transverse position. The usual traveling position may be different; however, account can be taken of the increased vehicle-edge distance in accordance with a minor correction to the method, which is given in Chap. 9. The method requires the following steps:

1. Using the values of α, θ, and μ [Eqs. (4.4), (4.5), and (4.6), respectively] which were used for ULS analysis, obtain D and C_f, for both external and internal portions, from the relevant chart of Fig. 4.13. For one-lane bridges the D and C_f values are the same as for ULS. Therefore, these values are obtained from Fig. 4.11a.

2. Obtain values of D_d for external and internal portions, using Eq. (4.7).

3. For each of the external and internal portions, obtain design live-load moments by multiplying the governing moments due to one line of wheels by $(S/D_d)(1 + DLA)$. It is noted that in the case of truck loadings the governing moment due to one line of wheels is the same as that calculated for the ULS analysis.

4. Take account of the increased vehicle-edge distance in accordance with the method of Chap. 9 since, for SLS I, the vehicle is positioned in the middle of the traveling lane.

Example 4.1 To illustrate the use of the method, a right noncomposite slab-on-girder bridge is appropriate; the analysis is for ULS, for evaluation of an existing bridge. The cross section of the bridge is shown in Fig. 4.14, and it is, of course, unnecessary to obtain initial values for initial proportioning. The calculations for this example are given in metric units only.

The first step is to determine live-load longitudinal moments, and in doing so we first disregard the possibility of composite action between the girders

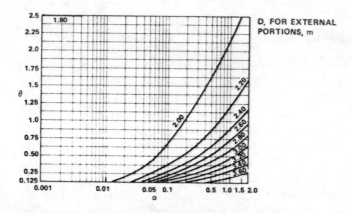

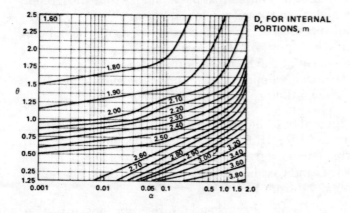

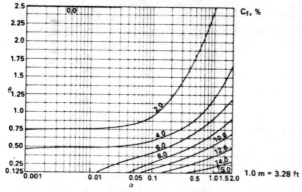

Figure 4.13a D and C_f charts for two-lane bridges, SLS I, one lane loaded.

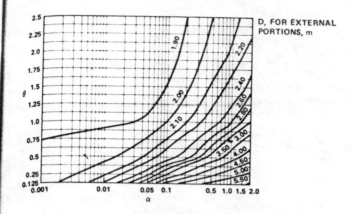

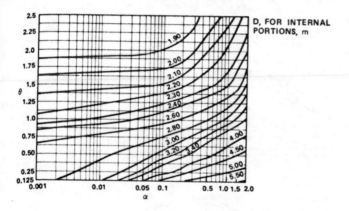

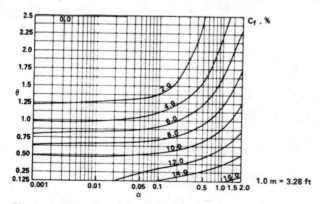

Figure 4.13b D and C_f charts for three-lane bridges, SLS I, one lane loaded.

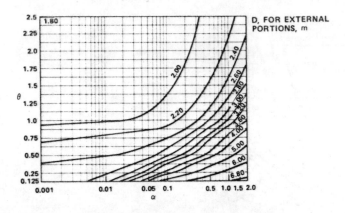

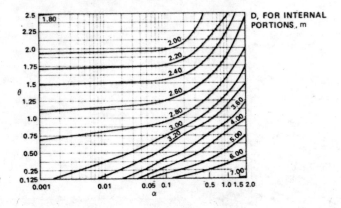

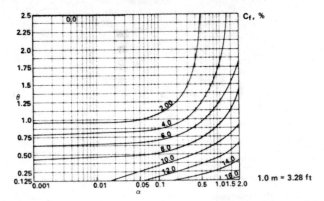

1.0 m = 3.28 ft

Figure 4.13c D and C_f charts for four-lane bridges, SLS I, one lane loaded.

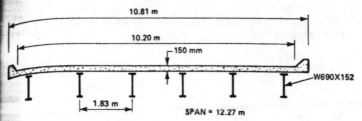

Figure 4.14   Cross section of a slab-on-girder bridge.

and the concrete deck slab. Relevant details of the bridge properties are as follows:

Span L = 12.27 m
Width $2b$ = 10.81 m
Girder spacing S = 1.83 m
Curb-to-curb width   = 10.20 m
Number of lanes (Table 4.1)   = 3
Design lane width   = 3.40 m
Slab thickness t = 0.15 m
Poisson's ratio   = 0.15
Second moment of area of a girder = 1.51×10^9 mm^4
Modular ratio n = 9.0

By using Eqs. (2.3), plate parameters for the noncomposite bridge, in millimeters, are found to be

$$D_x = \frac{nE_c(1.51 \times 10^9)}{1830} + E_c\frac{150^3}{12} = 7.71 \times 10^6 E_c$$

$$D_y = \frac{E_c(150^3)}{12(1 - 0.15^2)} \qquad = 0.29 \times 10^6 E_c$$

$$D_{xy} = G_c\frac{150^3}{6} \qquad\qquad = 0.24 \times 10^6 E_c$$

$$D_{yx} = D_{xy} \qquad\qquad\qquad = 0.24 \times 10^6 E_c$$

$$D_1 = 0.15(0.29 \times 10^6)E_c \qquad = 0.04 \times 10^6 E_c$$

$$D_2 = D_1 \qquad\qquad\qquad\quad = 0.04 \times 10^6 E_c$$

Using Eqs. (4.4), (4.5), and (4.6),

$$\alpha = \frac{(0.24 + 0.24 + 0.04 + 0.04) \times 10^6 E_c}{2(7.71 \times 0.29)^{0.5} \times 10^6 E_c} = 0.19$$

$$\theta = \frac{10.81}{2 \times 12.27}\left(\frac{7.71}{0.29}\right)^{0.25} \qquad = 1.00$$

$$\mu = \frac{3.40 - 3.30}{0.6} \qquad\qquad\qquad = 0.17$$

For external portions of this three-lane bridge, the chart of Fig. 4.11c is used, and yields a D value of 2.0 m. For internal portions the chart also gives a D value of 2.0 m, and the C_f value from the chart is found to be 7 percent. Therefore, for both external and internal girders,

$$D_d = 2.0\left(1 + \frac{0.17 \times 7}{100}\right) = 2.02 \text{ m}$$

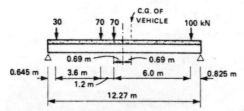

Figure 4.15 Longitudinal position of one line of wheels for maximum longitudinal moments.

In determining the first flexural frequency of the bridge, the action is taken to be composite and the frequency is found to be 9.5 Hz. Thence, from Fig. 4.9, the DLA is found to be 0.25.

The longitudinal position of the truck which produces the maximum bending moment in the span is shown in Fig. 4.15. The maximum moment due to one line of wheels in this position is readily found to be 425 kN · m. It is useful to check this value for accuracy by Fig. 4.7.

Hence, for both external and internal girders, the governing live-load moment for the ULS and SLS II is $(1.83/2.02)(1 + 0.25)(425) = 481$ kN · m. It should be noted that these moments are unfactored; i.e., no load factor has yet been applied to them.

It is instructive to investigate what the live-load moments will be if there is composite action between the girders and the concrete slab. Values of D_y, D_{xy}, D_{yx}, D_1, and D_2 remain the same as for the noncomposite structure. The value of D_x is derived from the combined second moment of area of a girder and its associated portion of the slab, which is calculated to be 32.93×10^9 mm⁴ in concrete units. Hence

$$D_x = \frac{32.93 \times 10^9}{1830} E_c = 17.99 \times 10^6 E_c$$

from which

$$\alpha = \frac{(0.24 + 0.24 + 0.04 + 0.04) \times 10^6 E_c}{2(17.99 \times 0.29)^{0.5} \times 10^6 E_c} = 0.12$$

$$\theta = \frac{10.81}{2 \times 12.27}\left(\frac{17.99}{0.29}\right)^{0.25} = 1.24$$

Charts of Fig. 4.11c for longitudinal moments in external and internal portions give D values of 1.89 m and 1.86 m, respectively, and a C_f value of 5.5

percent. Therefore, for external portions,

$$D_d = 1.89\left(1 + \frac{0.17 \times 5.5}{100}\right) = 1.91 \text{ m}$$

Similarly, for internal portions,

$$D_d = 1.86\left(1 + \frac{0.17 \times 5.5}{100}\right) = 1.88 \text{ m}$$

Since the first flexural frequency is always calculated by assuming the structure to act compositely, the DLA calculated for the earlier case is applicable in the present case also.

Using the same maximum moment due to one line of wheels as was calculated for the noncomposite case (i.e., 425 kN · m), we get the governing live-load moment for the ULS and SLS II, for the external portion, as

$$\frac{1.83}{1.91}(1 + 0.25)425 = 509 \text{ kN} \cdot \text{m}$$

Similarly, the governing moment for the internal portion is

$$\frac{1.83}{1.88}(1 + 0.25)425 = 517 \text{ kN} \cdot \text{m}$$

Hence the assumption of composite action between the girders and the slab results in an increase in live-load moments of 5.8 and 7.5 percent in the external and internal portions, respectively.

4.3   Method for AASHTO Loading

As mentioned in Sec. 4.1, the AASHTO specifications [1] provide "blanket" D values for different bridge types. For a given bridge type these values remain unaffected by the aspect ratio of bridges and details of their cross sections. Clearly, the AASHTO method is provided only as a design convenience. The specifications state that, in lieu of such an empirical design convenience, a rational analysis may be used, provided that the analysis is based on a theory accepted by the Committee on Bridges and Structures of the American Association of State Highway and Transportation Officials.

The method presented here is based upon the well-established orthotropic plate theory which is discussed in detail in Ref. 3. The orthotropic plate theory, which has been extensively validated with experimental results by several research workers, is also the basis for the current AASHTO method for slab-on-girder bridges [10]. Therefore, it is postulated that the orthotropic plate theory is already acceptable to the AASHTO bridge committee, and the use of the methods given in this book is compatible with the AASHTO specifications.

For consistency with the AASHTO specifications, both U.S. Customary and metric units are used in this section.

AASHTO Loading

The AASHTO design loadings are of three types:

1. A two-axle truck with lines of wheels which are 6.0 ft (1.83 m) apart.

2. A three-axle truck with lines of wheels which are 6.0 ft (1.83 m) apart.

3. A uniformly distributed lane load together with one knife-edge load for positive moment and shear calculations, and two knife-edge loads for negative moments in continuous spans.

All loadings have different levels, as shown in Fig. 4.16. The two-axle trucks and the corresponding lane loads are designated by H in the imperial system, or M in the metric system, followed by a number indicating the gross weight of the truck in tons (or tonnes). The three-axle trucks are designated by HS (or MS), followed by a number indicating the gross weight of the first two axles in tons (or tonnes).

As in the case of Ontario loading, the lane load governs only in longer spans. Maximum moments in simple spans due to one line of wheels, or half-lane loads, of some of the commonly used loadings, are shown in Fig. 4.17. It can be seen that differences in the H (M) and HS (MS) loadings of the same level occur only for spans between about 20 and 150 ft (6 and 46 m). For spans smaller than about 20 ft (6 m) and longer than about 150 ft (46 m), moments due to the two loadings are the same.

The distance between the two lines of wheels is nearly the same in the Ontario and AASHTO design loads. The same simplified methods of analysis could have been used for the two loadings, if it had not been for the differences in reduction factors due to simultaneous loading in adjacent lanes that exist between the two. It is noted that these reduction factors are implicit in the simplified methods for both Ontario and AASHTO loadings. The Ontario reduction factors for multilane loading are given in Sec. 4.2. For AASHTO loading, the reduction factors, as shown in Fig. 4.8b, are 1.0, 0.9, and 0.75 when two, three, and four lanes, respectively, are loaded.

Design Lanes

The AASHTO specifications require that bridges having roadway widths between 20 and 24 ft (6.096 and 7.315 m) have two equal design lanes.

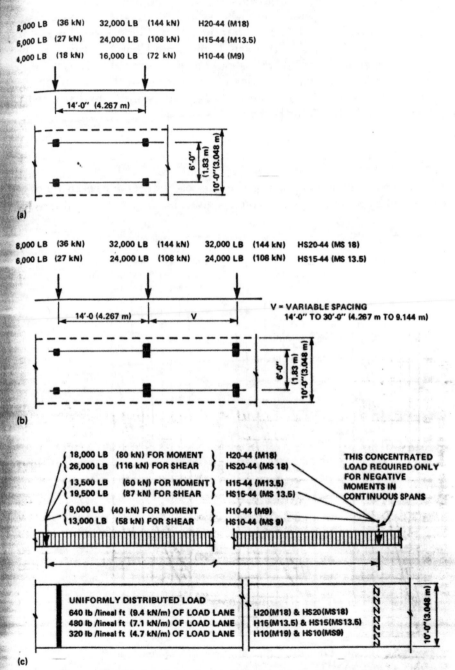

Figure 4.16 AASHTO design loads: (a) H (M) truck loading; (b) HS (MS) truck loading; (c) H (M) and HS (MS) lane loading.

Bridges with roadway widths outside this range are required to have 12-ft-wide (3.658-m-wide) lanes which should be arranged so as to result in maximum load effects. For the determination of longitudinal moments and longitudinal shears (which are discussed in Chap. 5), this requirement means that all design lanes should be packed as closely as possible toward one longitudinal free edge of the bridge, so that the fractional part of a lane, which is not required to be used, is positioned next to the other longitudinal free edge.

Impact

For superstructure components the live-load effects are required to be increased by an allowance I, which is given by

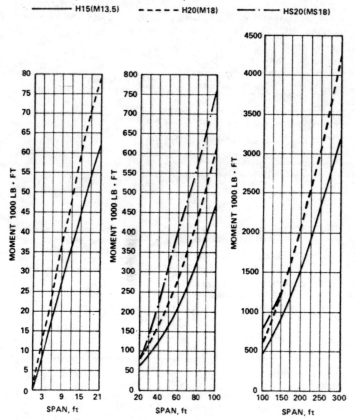

Figure 4.17 Moment due to one line of wheels or half-lane loading due to some AASHTO design loads.

$$I = \frac{50}{L + 125} \qquad \text{for U.S. Customary units}$$

$$(4.8)$$

$$\text{or} \quad I = \frac{15.24}{L + 38} \qquad \text{for metric units}$$

where L is the span in feet (or meters) of that portion of the bridge which is loaded to produce the maximum stress in the member under investigation. I is specified to have a maximum value of 0.3 and is equal to zero for timber bridges.

Simplified Method for Live-Load Moments

The method given here was developed for one-, two-, three-, and four-lane bridges. The governing load cases for bridges with different numbers of lanes are those which, after the application of the appropriate reduction factor for multilane loading, produce the maximum longitudinal moments. Governing load cases in bridges with different numbers of lanes and different values of α and θ are shown in Fig. 4.18.

The various steps in calculating live-load longitudinal moments are substantially the same as for Ontario loading, already described in Sec. 4.2. However, to avoid the necessity for constant reference back to that section, they are repeated here in full. For the design of new bridges all the steps should be followed. For the evaluation of an existing bridge, steps 1, 2, 4, and 5 are irrelevant and should be omitted.

1. Obtain an initial D_d value from Table 4.2 according to the bridge type and number of design lanes in the bridge.

2. Calculate the initial load fraction S/D_d, where $S =$ the actual girder spacing in the case of slab-on-girder bridges; or the spacing of webs in the case of voided slabs and cellular structures; or 1 ft (1 m) in the case of solid slabs, transversely prestressed laminated-wood bridges, and concrete-wood composite structures consisting of wood laminates and concrete overlays.

3. Treating the bridge as a one-dimensional beam, obtain bending moment diagrams due to the design loading as shown in Fig. 4.16.

4. Multiply the moments obtained in step 3 by $(S/D_d)(1 + I)$ to obtain the initial live-load moments, where the impact allowance I is obtained from Eq. (4.8). It is again emphasized that the reduction factor for multilane loading is implicit in the D_d values and should not be applied again.

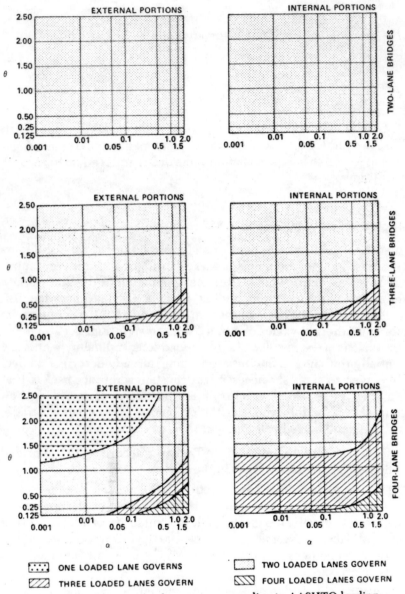

Figure 4.18 Governing load cases corresponding to AASHTO loading.

TABLE 4.2 Initial Longitudinal Moment D_d Values, in Meters (Feet), for ULS and SLS II of Ontario Code, and AASHTO Specifications

Bridge type	Number of design lanes			
	1	2	3	4
Slab bridges, voided slab bridges, and waffle slab bridges with nearly square waffles	1.75 (5.75)	2.00 (6.55)	2.25 (7.40)	2.50 (8.20)
Slab-on-girder bridges including steel girder with wood decking	1.90 (6.25)	1.90 (6.25)	2.05 (6.75)	2.15 (7.05)
Glulam girder bridges with wood decking	1.90 (6.25)	1.90 (6.25)	2.05 (6.75)	2.15 (7.05)
Sawn-timber stringer bridges with wood decking	1.80 (5.90)	1.80 (5.90)	1.85 (6.05)	1.90 (6.25)
Laterally prestressed laminated-wood bridges	1.90 (6.25)	1.90 (6.25)	2.05 (6.75)	2.15 (7.05)
Wood-concrete composite bridges	1.70 (5.60)	1.70 (5.60)	1.75 (5.75)	1.80 (5.90)

5. Assume that the live-load moments obtained in step 4 are sustained by a width S of the bridge for both external and internal portions, where S is as defined in step 2, and hence obtain initial proportions of the structure.

6. Calculate α and θ from Eqs. (4.4) and (4.5), respectively, using the procedure given in Sec. 2.2.

7. Calculate μ from one of the following expressions:

$$\mu = \frac{W_e - 11}{2} \quad \not> 1.0 \quad \text{for U.S. Customary units}$$

$$(4.9)$$

$$\text{or} \quad \mu = \frac{W_e - 3.3}{0.6} \quad \not> 1.0 \quad \text{for metric units}$$

where W_e is the design lane width (in feet for the U.S. Customary units expression and in meters for the metric units expression).

8. Corresponding to the values of α and θ, obtain values of D separately for external and internal portions, along with the value of C_f, from the relevant charts in Fig. 4.19a, b, and c for two-, three-, and four-lane

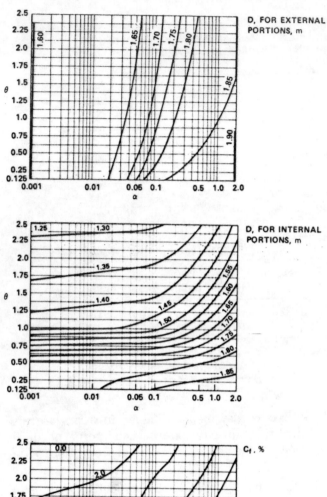

D, FOR EXTERNAL PORTIONS, m

D, FOR INTERNAL PORTIONS, m

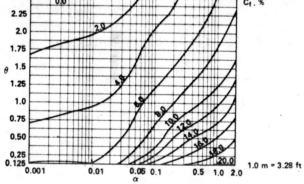

C_f, %

1.0 m = 3.28 ft

Figure 4.19a D and C_f charts for two-lane bridges, AASHTO loading.

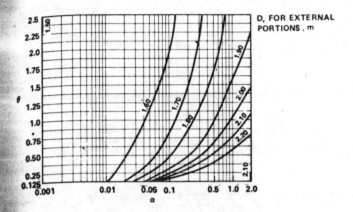

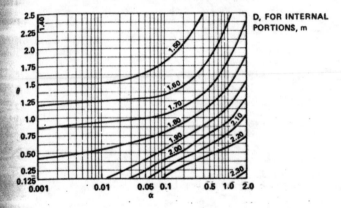

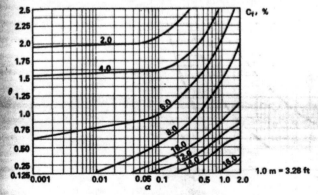

Figure 4.19b D and C_f charts for three-lane bridges, AASHTO loading.

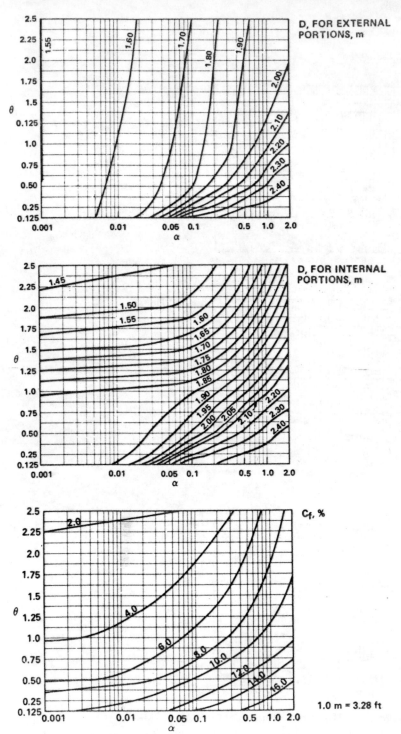

Figure 4.19c D and C_f charts for four-lane bridges, AASHTO loading.

bridges, respectively. For one-lane bridges obtain these values from Fig. 4.11a. It is noted that the charts provide values of D in meters. To obtain values in feet, divide these values by 0.3048.

9. Obtain the final value of D_d, separately for external and internal portions, from

$$D_d = D\left(1 + \frac{\mu C_f}{100}\right) \tag{4.10}$$

10. For each of the external and internal portions, obtain the final live-load design moments by multiplying the live-load moments due to one line of wheels or half-lane loads, as obtained in step 3 above, by $(S/D_d)(1 + I)$, where D_d has the value appropriate to the external or internal portion.

The summary of the above procedure is the same as for the Ontario loading and is given in Fig. 4.12.

The AASHTO specifications do not distinguish between loading for strength design and that for fatigue design. As pointed out in Ref. 9, this may result in grossly overconservative estimates of stresses for fatigue design. A more realistic requirement for fatigue would stipulate single-lane loading. An engineer wishing to analyze a bridge with a single lane loaded should use the SLS I method for Ontario loading, as given in Sec. 2.2. Since no reduction factor for multilane loading is involved, this method is equally applicable to AASHTO loads.

Example 4.2 A right slab-on-girder bridge with precast concrete girders and composite concrete deck slab is analyzed here by the above simplified method, for the HS 20-44 loading. Calculations are given in U.S. Customary units with metric units in parentheses. The cross section of the bridge is shown in Fig. 4.20. Relevant details are as follows:

Span L	= 85.0 ft (25.91 m)
Width $2b$	= 40.7 ft (12.41 m)
Girder spacing S	= 8.25 ft (2.51 m)
Roadway width	= 37.75 ft (11.51 m)
Number of lanes	= 3
Design lane width	= 12.0 ft (3.66 m)
Slab thickness t	= 7.5 in (190 mm)
Poisson's ratio	= 0.15
Modular ratio $\left(\dfrac{E,\text{ girder concrete}}{E,\text{ slab concrete}}\right)$	= 1.08

The bridge is analyzed by ignoring the barrier walls. The beneficial effect of such fairly heavy edge beams can be accounted for by using the method given in Chap. 9.

The combined second moment of area of a girder and the associated deck slab, in terms of deck slab concrete, is equal to 612,351 in⁴ (2.54×10^{11}

mm^4). The torsional inertia of a girder is calculated by approximating the cross section by a number of rectangles. Its value is found to be equal to 9952 in^4 (4.14 × 10^9 mm^4).

Using Eq. (2.4), the plate parameters for the composite bridge are calculated as follows, in inches, with E_c referring to slab concrete.

$$D_x = \frac{612,351}{8.25 \times 12} E_c \qquad\qquad\qquad = 6185E_c$$

$$D_y = \frac{7.5^3}{12} E_c \qquad\qquad\qquad\qquad = 35E_c$$

$$D_{xy} = \left(\frac{1.08 \times 9952}{8.25 \times 12} + \frac{7.5^3}{6}\right)G_c = 179G_c = 78E_c$$

$$D_{yx} = \frac{7.5^3}{6} G_c \qquad\qquad\qquad = 70G_c = 31E_c$$

$$D_1 = 0.15 \times 35E_c \qquad\qquad\qquad = 5E_c$$

$$D_2 = D_1 \qquad\qquad\qquad\qquad\qquad = 5E_c$$

It is noted that the ratio of the shear moduli for the girder and deck slab concrete remains the same as the ratio of the moduli of elasticity because the value of Poisson's ratio is assumed to be the same for the two concretes.

Using Eqs. (4.4), (4.5), and (4.9),

$$\alpha = \frac{(78 + 31 + 5 + 5)E_c}{2(6185 \times 35)^{0.5}E_c} = 0.13$$

$$\theta = \frac{40.7(12)}{2(85)(12)} \left(\frac{6185E_c}{35E_c}\right)^{0.25} = 0.87$$

$$\mu = \frac{12 - 11}{2} = 0.5$$

For the above values of α and θ the chart for external portions for three-

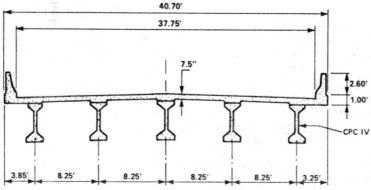

Figure 4.20 Cross section of a slab-on-girder bridge.

lane bridges given in Fig. 4.19b yields a D value of 1.73 m (5.68 ft). The chart for internal portions gives a D value of 1.78 m (5.84 ft), and the C_f value from the relevant chart is found to be 6.5 percent.

Hence, for external girders,

$$D_d = 5.68\left(1 + \frac{0.5 \times 6.5}{100}\right) = 5.86 \text{ ft } (1.79 \text{ m})$$

Similarly, for internal girders,

$$D_d = 5.84\left(1 + \frac{0.5 \times 6.5}{100}\right) = 6.03 \text{ ft } (1.84 \text{ m})$$

It is interesting to note that the value of D_d specified by AASHTO for this type of bridge is 5.5 ft (1.676 m). Thus the simplified analysis presented here would result in savings of 7 and 10 percent, respectively, in the live-load longitudinal moments for external and internal portions.

The impact factor corresponding to girders is obtained from Eq. (4.8):

$$I = \frac{50}{85 + 125} = 0.238$$

By employing the usual method, the maximum moment due to one line of wheels or half the lane load of HS 20 loading, for an 85-ft (25.91-m) span is found to be 627,000 lb·ft (850 kN·m), and Fig. 4.17 can be used for confirming the accuracy of this calculated maximum moment. Hence the governing live-load longitudinal moment in an external girder is

$$\frac{8.25}{5.86}(1 + 0.238)(627,000) = 1,092,810 \text{ lb·ft } (1482 \text{ kN·m})$$

Similarly, the corresponding moment in an inside girder is

$$\frac{8.25}{6.03}(1 + 0.238)(627,000) = 1,062,000 \text{ lb·ft } (1442 \text{ kN·m})$$

References

1. American Association of State Highway and Transportation Officials (AASHTO): *Standard Specifications for Highway Bridges*, Washington, D.C., 1977.
2. Bakht, B., Cheung, M. S., and Aziz, T. S.: The application of a simplified method of calculating longitudinal moments to the proposed Ontario highway bridge design code, *Canadian Journal of Civil Engineering*, 6(1), 1979, pp. 36–50.
3. Cusens, A. R., and Pama, R. P.: *Bridge Deck Analysis*. Wiley, London, 1975.
4. Dorton, R. A., and Csagoly, P. F.: *The Development of the Ontario Bridge Code*, Ministry of Transportation and Communications, Downsview, Ontario, Canada, 1977.
5. Hendry, A. W., and Jaeger, L. G.: *The Analysis of Grid Frameworks and Related Structures*, Prentice-Hall, Englewood Cliffs, N.J., 1959.
6. Ministry of Transportation and Communications: *Ontario Highway Bridge Design Code (OHBDC)*, 1st ed., Downsview, Ontario, Canada, 1979.
7. Ministry of Transportation and Communications: *Ontario Highway Bridge Design Code (OHBDC)*, 2d ed., Downsview, Ontario, Canada, 1983.

8. Morice, P. B., and Little, G.: *The Analysis of Right Bridge Decks Subjected to Abnormal Loading*, Cement and Concrete Association's Report Db11, London, 1956.
9. Radkowski, A. F., Bakht, B., and Billing, J. R.: Design and testing of a 125 m span plate girder bridge, *International Conference on Short and Medium Span Bridges*, Toronto, Ontario, Canada, August 1982.
10. Sanders, W. W., Jr., and Elleby, H. A.: *Distribution of Wheel Loads on Highway Bridges*, National Co-operative Highway Research Program Report 83, Washington, D.C., 1970.

METHODS FOR LONGITUDINAL SHEARS IN SHALLOW SUPERSTRUCTURES

5.1 Introduction

It was noted in Sec. 3.3 that the transverse distribution of longitudinal shears is more localized than that of longitudinal moments. In Sec. 4.1, it was further noted that the basis of the simplified method for longitudinal moments is the assumption that the transverse distribution pattern of these moments is substantially the same from one transverse section to another. There is, at first sight, an apparent conflict between these two statements, as discussed below. Closer scrutiny thereafter removes the anomaly.

Differences between Moment and Shear Distributions

Consider a right bridge idealized as an orthotropic plate and subjected to a load as shown in Fig. 5.1. Further, consider two transverse sections, identified as 1-1 and 2-2 in Fig. 5.1, away from the load and at a distance Δx apart. Let the total longitudinal moments along 1-1 and 2-2 be equal to M_1 and M_2, respectively. The total longitudinal shear V_0 at section 0-0,

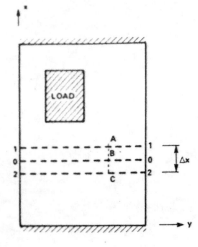

Figure 5.1 Plan of a bridge idealized by an orthotropic plate.

which is midway between 1-1 and 2-2, is given by

$$V_0 = \frac{M_1 - M_2}{\Delta x} \tag{5.1}$$

It is recalled that the above equation is the finite difference equivalent of

$$V_x = \frac{\partial M_x}{\partial x}$$

Now, consider points A and C at sections 1-1 and 2-2, respectively, on the same longitudinal line, and let the longitudinal moment intensities at these two points be designated by M_{xA} and M_{xC}. If M_{xA} is given by the following expression,

$$M_{xA} = KM_1 \tag{5.2}$$

where K is a constant, then at point C, according to the premise that the transverse distribution of M_x is not affected by the longitudinal position of the reference point, M_{xC} should be given by:

$$M_{xC} = KM_2 \tag{5.3}$$

If the effects of twisting moments on longitudinal vertical planes are ignored, the intensity of longitudinal shear V_{xB} at point B, which is midway between A and C, should be given by:

$$V_{xB} = \frac{M_{xA} - M_{xC}}{\Delta x} \tag{5.4}$$

Substituting the expressions for M_{xA} and M_{xC} from Eqs. (5.2) and (5.3), respectively, Eq. (5.4) can be rewritten as

$$V_{xB} = K\left(\frac{M_1 - M_2}{\Delta x}\right) \tag{5.5}$$

or, from Eq. (5.1),

$$V_{xB} = K V_0 \tag{5.6}$$

Equation (5.6) implies that longitudinal shear has the same transverse pattern of distribution as longitudinal moments, so that the assertion that the patterns are different, made in Sec. 3.3, appears to be false.

The apparent paradox in the argument emerges from the fact that the pattern of transverse distribution of M_x is not *exactly* independent of the longitudinal position of the reference point; if it were, then indeed M_x and V_x would have the same distribution patterns and could be analyzed by the same simplified methods. The slight dependence of K on the longitudinal position of the reference point has a relatively small effect on the accuracy of the simplified methods for analyzing longitudinal moments. However, when this slight dependence is ignored in the process of differentiation of moments to obtain shears, as is done in obtaining Eq. (5.5), the values of shears obtained in this way depart from the true values by a much wider margin. In fact, the transverse distribution of shears is significantly different from that of moments.

This argument is illustrated with the help of Table 5.1, which contains some results of the grillage analysis of a slab-on-girder bridge under a vehicle load. From a scrutiny of results in rows 1 and 2 of Table 5.1, it can be seen that the distribution coefficients for longitudinal moments in girders at two different transverse sections are not exactly the same but are fairly close. However, the coefficients for longitudinal shear midway between the above two sections, shown in row 4, are significantly different from the corresponding moment coefficients. Longitudinal shears obtained from longitudinal moments according to Eq. (5.4) are given in row 5. It can be seen that these values and their distribution coefficients are of the same order of magnitude as those of shears obtained from grillage analysis (row 4).

It is instructive at this point to calculate longitudinal moments and their coefficients at section 2-2 (identified in the figure in Table 5.1) by means of the average moment at section 1-1 and the distribution coefficients for section 2-2. It can be seen that the moments obtained in this way, which are listed in row 3, are not significantly different from the actual moments. However, when these moments are used to calculate longitudinal shears according to Eq. (5.4), the results, given in row 6, are

TABLE 5.1 Longitudinal Moments and Shears from Grillage Analysis of a Bridge

	GIRDER NO.		
	①	⑤	⑨
1 M_{x2} along section 2-2 1000 lb/in	898	1,410	74
Distribution coefficient (Average moment = 989,000 lb/in)	0.91	1.43	0.07
2 M_{x1} along section 1-1 1000 lb/in	883	1,216	81
Distribution coefficient (Average moment = 896,000 lb/in)	0.98	1.36	0.09
3 M'_{x1} along section 1 - 1 using distribution coefficient for section 2-2	815	1,281	63
4 Actual V_{x0} at midspan, kips	0.22	4.40	-0.24
Distribution coefficient (Average shear = 1930 lb)	0.11	2.28	-0.12
5 V_{x0} obtained from: $V_{x0} = \dfrac{M_{x2} - M_{x1}}{48''}$	0.31	4.04	-0.15
Distribution coefficient (Average shear = 1930 lb)	0.16	2.09	-0.08
6 V_{x0} obtained from $V_{x0} = \dfrac{M_{x2} - M'_{x1}}{48''}$	1.75	2.92	0.23
Distribution coefficient (Average shear = 1930 lb)	0.91	1.51	0.12

substantially different from the actual values of shears (row 4), thus demonstrating that Eqs. (5.2) and (5.3) should be written as

$$M_{xA} \simeq KM_1$$
$$M_{xB} \simeq KM_2 \tag{5.7}$$

In this case Eqs. (5.4) and (5.6) become false, indicating that moments and shears can have significantly different transverse distributions.

Developmental Background

The methodology for developing the simplified method given in this chapter was conceived by the authors for the 1983 edition of the *Ontario Highway Bridge Design Code* [7]. Relevant details of the developmental background, which are not yet reported in the literature, are given here.

The format of the method is the same as that for longitudinal moments (Chap. 4); that is, a D-type approach is used. The values of D for longitudinal shears were obtained by the grillage analogy method [5] instead of by orthotropic plate analysis, which was used for determining D values for longitudinal moments. It is recalled that the former analysis requires idealization by discrete beam members, and the latter by a continuous medium. If the details of the cross section of a bridge are known, then in grillage analysis it is customary to have the conceptual grillage reflect as closely as possible the actual bridge; this is achieved, for example, by having the positions of the longitudinal grillage beams coincide with the positions of actual girders. Such details of the particular cross section are not known when a general method is developed for different bridge cross sections and types. For grillage analysis the idealization was done in two steps: In the first step the bridge was idealized as an orthotropic plate; in the second step this orthotropic plate was idealized by a grillage comprising nine equally spaced longitudinal beams and seven equally spaced transverse beams, as shown in Fig. 5.2.

The grillage analysis was chosen over the orthotropic plate analysis because of some ambiguity over the physical relevance of longitudinal shear in orthotropic plates. Two types of shear forces are defined in Ref. 3, namely, longitudinal shearing force Q_x and longitudinal supplemented

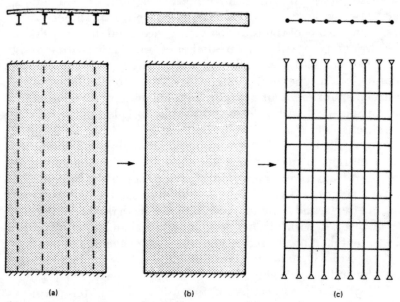

Figure 5.2 Idealization of a bridge by a grillage through an orthotropic plate. (*a*) Actual bridge. (*b*) Orthotropic plate. (*c*) Grillage.

shearing force V_x. These two forces are expressed in terms of the deflection w as follows:

$$Q_x = \left[D_x \frac{\partial^3 w}{\partial x^3} + (D_{yx} + D_1) \frac{\partial^3 w}{\partial x \, \partial y^2} \right] \tag{5.8}$$

$$V_x = Q_x - D_{xy} \frac{\partial^3 w}{\partial x \, \partial y^2} \tag{5.9}$$

V_x is also referred to as the Kirchhoff's edge reaction (see Ref. 4). Some authors (e.g., see Ref. 2) argue that this force effect is only a mathematical abstraction necessary to solve the small deflection plate bending problem, and that it has no physical significance. Others (e.g., see Ref. 6) maintain that the Kirchhoff's edge reaction is a close approximation of the actual. The Kirchhoff's edge reaction is a combination of shearing force Q_x and rate of change of twisting moment, $\partial M_{xy}/\partial y$.

Shear in a grillage is a readily identifiable force effect. Therefore, it was decided to resort to grillage analysis for obtaining values corresponding to longitudinal shear.

The limitations on bridge geometry for the application of the method relate to the realistic idealization of the bridge as an orthotropic plate. Therefore, these limitations are the same as those that apply to the method of determining longitudinal moments, as given in Sec. 4.1.

As discussed in Sec. 1.8, the distribution of longitudinal shear is characterized by two parameters: α_1 and θ [Eqs. (1.3) and (1.5), respectively]. However, from results of rigorous analyses it was found that the D value for a given bridge type with a given number of lanes varies within only a relatively small range. It was confirmed that specifying a single value for bridges of a certain type and having a specific number of lanes would involve a maximum error of ±5 percent. This finding, which is also discussed in Sec. 10.4 with reference to Table 10.1, led to the method given in this chapter.

Results of grillage analysis were interpolated to obtain D values corresponding to a beam spacing of 2.0 m (6.56 ft), except for concrete-wood composite and transversely post-tensioned laminated bridges. For these latter bridges, a 1-m (3.28-ft) width was used as the basis for averaging of effects, and D values were obtained from these averages. In the case of bridges with girders spaced at less than 2.0 m (3.28 ft), account can be taken of the actual value of girder spacing by means of Eqs. (5.10) and (5.11), given in Secs. 5.2 and 5.3.

A careful scrutiny of D values for longitudinal shear as given in Tables 5.2 and 5.3 will reveal some apparent inconsistencies in their patterns. For example, examination of the D values for slab-on-girder bridges as given in Table 5.3 shows that the value decreases when the number of

TABLE 5.2 Shear _D_ Values, in Meters, Corresponding to Ontario Loading

Limit state	Bridge type	Number of lanes in the bridge			
		1	2	3	4
SLS I	Slab and voided slab bridges	2.05	2.10	2.25	2.50
	Slab-on-girder bridges°	1.75	1.80	1.90	2.10
	Transverse laminated deck on sawn-stringer or glulam girders°	1.75	1.80	1.90	2.10
	Concrete deck slab on longitudinally laminated-wood deck	1.85	1.95	2.10	2.30
	Prestressed laminated-wood decks	1.65	1.75	1.85	2.00
ULS and SLS II	Slab and voided slab bridges	2.05	1.95	1.95	2.15
	Slab-on-girder bridges°	1.75	1.70	1.85	1.90
	Transverse laminated deck on sawn-stringer or glulam girders°	1.75	1.70	1.85	1.90
	Concrete deck slab on longitudinally laminated-wood deck	1.85	1.70	1.85	1.90
	Prestressed laminated-wood decks	1.65	1.70	1.80	1.85

° When S is smaller than 2.0 m, use Eq. (5.10).

TABLE 5.3 Shear _D_ Values, in Meters (Feet), Corresponding to AASHTO Loading

Bridge type	Number of lanes in the bridge			
	1	2	3	4
Slab and voided slab bridges	2.05 (6.75)	1.90 (6.25)	2.00 (6.55)	2.25 (7.40)
Slab-on-girder bridges°	1.75 (5.75)	1.60 (5.25)	1.70 (5.60)	1.90 (6.25)
Transverse laminated deck on sawn-timber or glulam girders°	1.75 (5.75)	1.60 (5.25)	1.70 (5.60)	1.90 (6.25)
Concrete deck on longitudinally laminated-wood deck	1.85 (6.05)	1.75 (5.75)	1.90 (6.25)	2.10 (6.90)
Prestressed laminated-wood decks	1.65 (5.40)	1.45 (5.25)	1.65 (5.40)	1.55 (5.90)

° When S is smaller than 2.0 m, use Eq. (5.10) for metric units calculation and Eq. (5.11) for U.S. Customary units calculation.

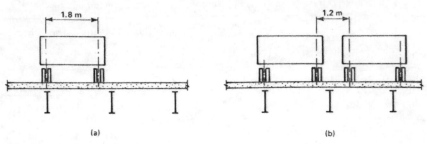

Figure 5.3 Transverse vehicle positions: (*a*) single vehicle; (*b*) two closely spaced vehicles.

lanes increases from one to two and then rises again when the number of lanes goes to three. The explanation for this behavior is instructive and now follows.

After taking the reduction factors for multilane loading into account, the governing D values in most two-, three-, and four-lane bridges for the case mentioned above arise from the "two-lane-loaded" case. A single vehicle has two lines of wheels which are 1.8 m (5.91 ft) apart; when two vehicles are placed side by side transversely, as close as possible, the adjacent lines of wheels of the two vehicles are only 1.2 m (3.94 ft) apart, as shown in Fig. 5.3. These two lines of wheels result in a more intense concentration of V_z than the lines of wheels (1.8 m, or 5.91 ft, apart) of a single vehicle — hence the decrease in D value when the number of lanes is increased from one to two. Any further increase in the number of lanes merely increases the width of the bridge and reduces the modification factor for multilane loading, as a result of which more and more portions of the transverse cross section of the bridge become available for sharing the load, with a consequent increase in the D value.

Maximum longitudinal shears due to vehicle loads are not as sensitive to the transverse position of the vehicles as are longitudinal moments. Consequently, the external and internal portions of the bridge (as defined in Sec. 4.1) are subjected to more or less the same maximum longitudinal shears. Accordingly, for the calculation of governing live-load longitudinal shears, no distinction is made between external and internal portions, and the results of the simplified analysis are applicable to the entire cross section.

5.2 Methods for Ontario Loading

Loadings, dynamic load allowance (DLA), and lane widths pertaining to the Ontario highway bridge design code [7] are discussed in some detail in Sec. 4.2. Governing simple beam shears due to one line of wheels or one half-lane loading are given in Fig. 5.4.

Simplified Method for ULS and SLS II

Using the reduction factors for multilane loading, the number of loaded lanes which govern the design for live-load longitudinal shears are shown in Table 5.4.

The various steps in calculating live-load longitudinal shears in shallow superstructures are as follows:

1. Select a D value corresponding to ULS and SLS II from Table 5.2 in accordance with the type of bridge and the number of design lanes in the bridge. When there are more than four lanes in the bridge, use values for four-lane bridges. In the case of bridges with girders that are spaced at less than 2.0 m apart, the following expression should be used for calcu-

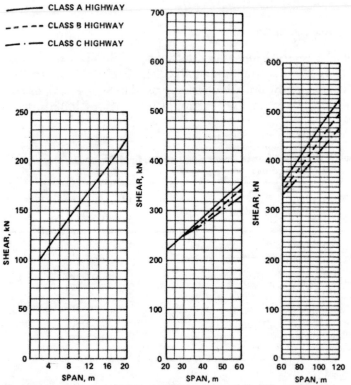

Figure 5.4 Shears due to one line of wheels or half-lane loading, OHBDC.

lating D_d:

$$D_d = \left(\frac{S}{2.0}\right)^{0.25} D \tag{5.10}$$

where S is in meters. For all other cases D_d should be taken to be the same as D.

2. Calculate the load fraction S/D_d, where $S =$ the actual girder spacing in the case of slab-on-girder bridges; or the spacing of webs in the case of voided slabs and cellular structures; or 1 m in the case of solid slabs, transversely prestressed laminated-wood bridges, and concrete-wood composite bridges composed of wood laminates and concrete overlays.

3. Treating the bridge as a one-dimensional beam, obtain shear force diagrams due to one line of wheels of the truck or one-half of the lane loading; the maximum values for both are shown in Fig. 5.4.

4. Multiply the shears obtained in step 3 above by (S/D_d) $(1 + DLA)$ to obtain the design value for longitudinal shear. This value is applicable for both external and internal portions.

Simplified Method for SLS I

The method for SLS I is the same as that for ULS and SLS II, except that the D values are read corresponding to SLS I in Table 5.2. Increased vehicle-edge distance for SLS I, which is discussed in Sec. 4.2, does not have any significant effect on the maximum intensity of longitudinal shear.

TABLE 5.4 Number of Loaded Lanes Governing Longitudinal Shear D Values for Both Ontario and AASHTO Loadings

Bridge type	Number of lanes in the bridge			
	1	2	3	4
Slab and voided slab bridges	1	2	2	3
Slab-on-girder bridges	1	2	2	2
Transverse laminated deck on sawn-timber or glulam girders	1	2	2	2
Concrete deck slab on longitudinally laminated-wood deck	1	2	2	2
Prestressed laminated-wood decks	1	2	2	2

Example 5.1 The same bridge that was analyzed in Example 4.1, the cross section of which is shown in Fig. 4.14, is analyzed here for SLS I, for longitudinal shear at quarter span.

The slab-on-girder bridge has three lanes. From Table 5.2, the value of D for SLS I is found to be 1.90 m (6.25 ft). The girder spacing S is 1.83 m (6.00 ft). Hence from Eq. (5.10)

$$D_d = \left(\frac{1.83}{2.0}\right)^{0.25} (1.90) = 1.86 \text{ m } (6.10 \text{ ft})$$

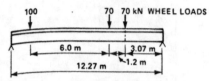

Figure 5.5 Wheel positions for maximum shear at quarter span.

The longitudinal position of the vehicle to induce maximum beam shear at quarter span is shown in Fig. 5.5. Maximum shear due to one line of wheels at this section is equal to 114 kN (25,627 lb).

The DLA, calculated as in Sec. 4.2, is 0.25. Hence the maximum longitudinal shear per girder at quarter span, due to live load, is $(1.83/1.86)(1 + 0.25)(114) = 140$ kN (31,472 lb). It should be noted that this value of longitudinal shear is unfactored; i.e., a load factor has not been applied to it.

5.3 Method for AASHTO Loading

Details of the AASHTO loading, design lanes, and impact factors are given in Sec. 4.3. Maximum simple span shears due to some AASHTO loadings are given in Fig. 5.6. The numbers of loaded lanes which govern the longitudinal shear D values, in bridges with various numbers of lanes, are the same as for Ontario loading and are shown in Table 5.4.

The method for determining live-load longitudinal shears at any transverse section due to AASHTO loadings is the same as the simplified method for ULS and SLS II for Ontario loading, as presented in Sec. 5.2, except that the D values for shears are selected from Table 5.3. When U.S. Customary units are used, the following equation should be used instead of Eq. (5.10).

$$D_d = \left(\frac{S}{6.56}\right)^{0.25} D \tag{5.11}$$

where S is in feet.

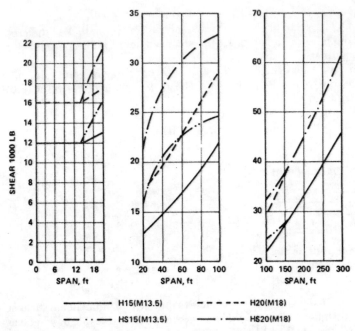

————————	H15(M13.5)	— — — — — H20(M18)
— ·· — ·· —	HS15(M13.5)	— · — · — HS20(M18)

Figure 5.6 Shear due to one line of wheels or half-lane loading due to AASHTO design loads.

Example 5.2 The same bridge that was analyzed for longitudinal moments in Example 4.2 is analyzed for maximum longitudinal shear at the support due to the HS 20-44 loading. The cross section of the three-lane slab-on-girder bridge is shown in Fig. 4.20.

From Table 5.3 the relevant D value is 5.60 ft (1.70 m). The maximum longitudinal shear due to one line of wheels of the HS 20-44 truck in a bridge of 85-ft span is 32,100 lb. It is noted that, for this span, the truck, rather than the lane, loading produces the maximum shear. The impact factor, as calculated in Sec. 4.3, is 0.238. Hence the maximum longitudinal shear sustained by each girder is given by $(8.25/5.60)(1 + 0.238)(32,100) = 58,545$ lb (260 kN).

References

1. American Association of State Highway and Transportation Officials (AASHTO): *Standard Specifications for Highway Bridges*, Washington, D.C., 1977.
2. Coull, A.: Corner forces in plate theory, *The Engineer*, vol. 220, 1965, p. 335.
3. Cusens, A. R., and Pama, R. P.: *Bridge Deck Analysis*, Wiley, London, 1975.
4. Jaeger, L. G.: *Elementary Theory of Elastic Plates*, Pergamon Press, Oxford, England, 1964.

5. Jaeger, L. G., and Bakht, B.: The grillage analogy in bridge analysis, *Canadian Journal of Civil Engineering*, 9(2), 1982, pp. 224–235.
6. Merrifield, B. C.: Reply to discussion on "Abutment reactions in right bridge decks," *Concrete*, May 1969, p. 175.
7. Ministry of Transportation and Communications: *Ontario Highway Bridge Design Code* (*OHBDC*), 2d ed., Downsview, Ontario, Canada, 1983.

6

METHODS FOR
TRANSVERSE MOMENTS
IN SHALLOW
SUPERSTRUCTURES

6.1 Introduction

In this chapter, two types of methods are given for calculating transverse moments. One type is developed from the results of linear elastic analysis; the other, which is applicable only to concrete deck slabs of slab-on-girder bridges, is based upon a well-tested empirical approach. The latter type of method takes account of the arching effect in slabs, which is discussed in Sec. 3.4 and which significantly enhances their load-carrying capacity.

Variation of Global Moments

As discussed in Sec. 3.4, global transverse moments in slab-on-girder bridges correspond to the overall transverse flexure of the bridge, with local bending effects between girders being neglected. In slab bridges, since there are no girders present to cause local bending effects, the total transverse moments can be likened to the global transverse moments of slab-on-girder bridges. Hence, the comments given below on the variation of global transverse moments are also applicable to the total transverse moments in slab bridges.

In a right single-span bridge, the maximum global transverse moment on a cross section occurs at or near the bridge centerline, as shown in Fig. 6.1. The transverse moment falls to zero at the longitudinal free edges. The variation of the global transverse moment in the transverse direction from maximum in the middle to zero at the edges is given approximately by the following expression:

$$M_{yg} = M_{yG} \cos \frac{\pi y}{2b} \tag{6.1}$$

where M_{yg} is the intensity of the global transverse moment and M_{yG} is the maximum intensity of the global transverse moment (i.e., the maximum value of M_{yg}), as shown in Fig. 6.1. It should be emphasized that Eq. (6.1) is only an approximate, usually upper-bound, expression for global transverse moments away from the bridge centerline. It was arrived at by studying the results of orthotropic plate analysis [4] and grillage analysis [5] for a large number of slab and slab-on-girder bridges. In the case of slab-on-girder bridges, the transverse distance between external girders should be used instead of $2b$ in Eq. (6.1).

The magnitude of the maximum transverse global moments remains almost constant within the middle two-thirds of the longitudinal span of the bridge, as shown in Fig. 6.1. The global transverse moment at the supports, because of zero transverse curvature, is equal to zero. (By contrast, it should be noted that in slab-on-girder bridges the local transverse moment at supports may not be zero.) Variation of the global moment in the outer one-sixth of the span, i.e., between zero and maximum, is of the form shown in Fig. 6.1.

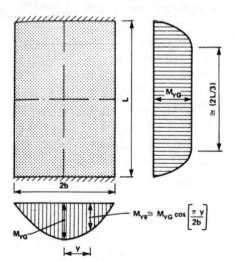

Figure 6.1 Variation of maximum transverse global moments in a bridge.

Idealization for Linear Elastic Analysis

The analytical method for determining global transverse moments in slab-on-girder bridges, as given below in Sec. 6.3, was developed from the results of grillage analysis. For all cases an 8×8 grillage mesh was employed, using the idealization process shown in Fig. 5.2.

A one-to-one relationship between actual girders of the bridge and longitudinal beams of the grillage may not be necessary when responses such as longitudinal moments and shears are investigated. However, for the determination of global transverse moments, the absence of this one-to-one relationship does have an adverse effect on the accuracy of results. If the idealized structure, i.e., the grillage, has more longitudinal beams than the actual number of girders in the bridge, then the analysis will predict a higher value of global transverse moments than actually occurs. Because of this, it should be noted that when there are fewer than nine girders in the bridge, the values of global transverse moments, as predicted by the method of Sec. 6.3, will be on the conservative side. The degree of conservatism may be as high as 20 percent in the extreme case of a bridge wth only four girders. It is recalled that the methods given in this chapter are not applicable to slab-on-girder bridges with fewer than four girders.

Increase of the actual number of girders in the bridge beyond nine has a relatively small effect on the transverse moments.

6.2   Empirical Method for Slab-on-Girder Bridges

Design of New Bridges

The Ontario highway bridge design code [3] requires that the deck slabs of new slab-on-girder bridges which conform to certain conditions shall be designed according to an empirical method which is specified in the code. According to this method, the slab thickness should be one-fifteenth of the deck slab span, with a minimum of 225 mm (8.86 in), and the reinforcement should comprise two orthogonal meshes with a minimum area of reinforcement, in each direction of each mesh, of 0.3 percent of the concrete area. The empirical design is applicable only when the following conditions are met:

1. The slab concrete strength is at least 30 MPa (4350 lb/in²).

2. The slab span does not exceed 3.7 m (12.14 ft).

3. The slab extends at least 1.0 m (3.3 ft) beyond exterior girders, or has a curb of equivalent area of cross section.

4. The spacing of reinforcing bars does not exceed 300 mm (11.81 in).

5. The bridge, if it has steel I or box girders, has diaphragms or cross frames extending throughout the cross section between external girders, at a maximum spacing of 8.0 m (26.25 ft).

6. The bridge has diaphragms or cross frames at all supports.

The requirement of a minimum slab thickness of 225 mm (8.86 in) is not related to the strength of the slab, which is governed predominantly by the span-to-thickness ratio of the slab, but rather to considerations of durability. It is believed that slabs and other concrete components exposed to deicing salts should have reinforcement with a minimum cover of 50 mm (1.97 in) from the salt-exposed surface. A survey of depths of cover in slabs, conducted in Ontario, showed that the standard deviation of the depth of cover is about 10 mm (0.39 in). Hence, to ensure that in 97.5 percent of cases the actual depth of cover would be at least 50 mm (1.97 in), a depth of cover of 70 mm (2.76 in) has been specified in the code. A 40-mm (1.57-in) depth of cover is specified for the bottom face. These requirements, together with the requirement of a minimum spacing of 25 mm (0.98 in) between the top and bottom layers of reinforcement, led to an overall minimum deck slab thickness of 225 mm (8.86 in).

In the case of skew bridges, the deck slab span is required to be measured along the skew direction. For bridges having skew angles greater than 20°, the reinforcement requirement in the end portions identified in Fig. 6.2 is increased to 0.6 percent isotropic reinforcement, as compared with 0.3 percent isotropic reinforcement in right or nearly right bridges.

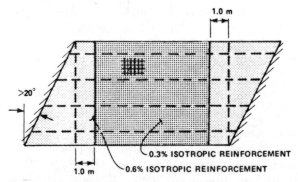

Figure 6.2 Reinforcement requirements in deck slab of skew slab-on-girder bridges.

Evaluation of Existing Bridges

For evaluation of the load-carrying capacity of existing bridges, the Ontario code provides charts for the unfactored resistance of deck slabs having different thicknesses, concrete strengths, span lengths, and reinforcement ratios. The resistance is given in terms of the intensity of concentrated loads which would cause failure in the slab. These charts, which are based on the findings reported in Refs. 2 and 3, are reproduced in Fig. 6.3. In the charts, f'_c is the compressive strength of concrete in megapascals (the stress in pounds per square inch is obtained by multiplying the megapascal value by 145), R_n is the unfactored resistance of the deck slab in kilonewtons (the equivalent resistance in pounds-force is obtained by multiplying the kilonewton values by 225), and q is the percentage tensile reinforcement at mid-span. The minimum thickness of competent concrete in a deteriorated deck slab is taken as the slab thickness. It should be noted that the slab resistances given in these charts are unfactored resistances. For the concrete deck slabs of slab-on-girder bridges, the Ontario code specifies a dynamic load allowance of 0.40, a performance factor ϕ of 0.5, and a load factor of 1.4. Using these values, the minimum slab resistance required to safely sustain a 70-kN wheel load is found to be 274 kN.

When the above method is used, the Ontario code does not require the calculation of transverse moments.

6.3 Analytical Method for Slab Bridges

The method given in this section was developed for the Ontario code by the authors. Orthotropic plate theory [4] was employed to develop the method. For the method to be applicable, the structure should be capable of being idealized as an isotropic plate, and therefore should conform to the limitations listed in Sec. 4.1.

It was found that the maximum transverse moments in slab bridges were induced when only a single lane, located at the middle of the bridge, was loaded. Since no reduction factors for multilane loading are involved in this case, the method is equally applicable to both Ontario and AASHTO loadings.

The various steps involved in the application of the method are as follows:

1. Calculate the maximum moment due to one-lane loading by treating the bridge as a one-dimensional beam. This moment corresponds to twice the moments given in Figs. 4.7 and 4.17 for Ontario and AASHTO loads, respectively. Divide this maximum moment by the bridge width in order to obtain the average longitudinal moment $M_{x,av}$.

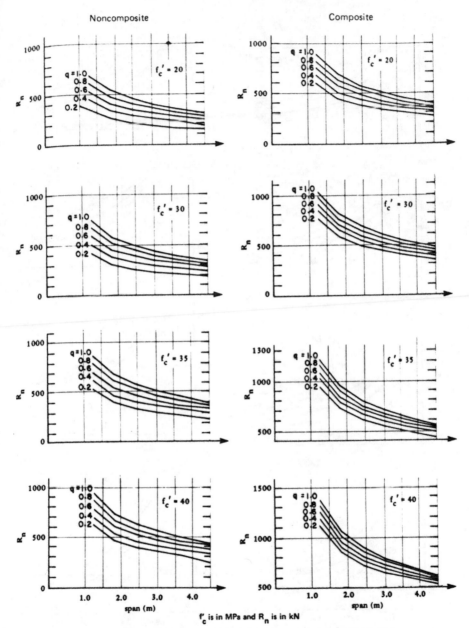

Figure 6.3a Charts for ultimate strengths of 150-mm concrete deck slabs.

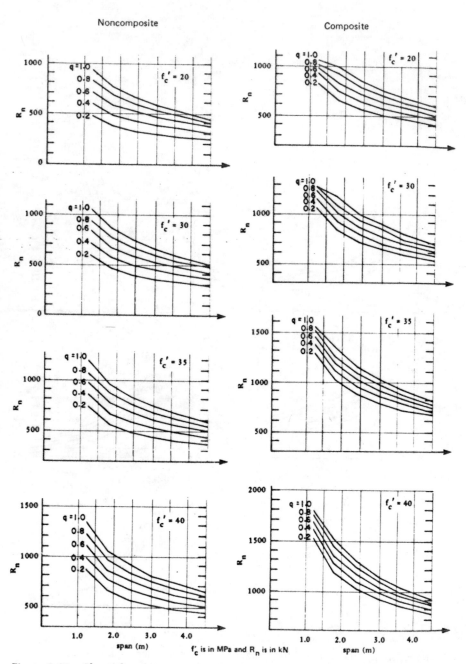

Figure 6.3b Charts for ultimate strengths of 175-mm concrete deck slabs.

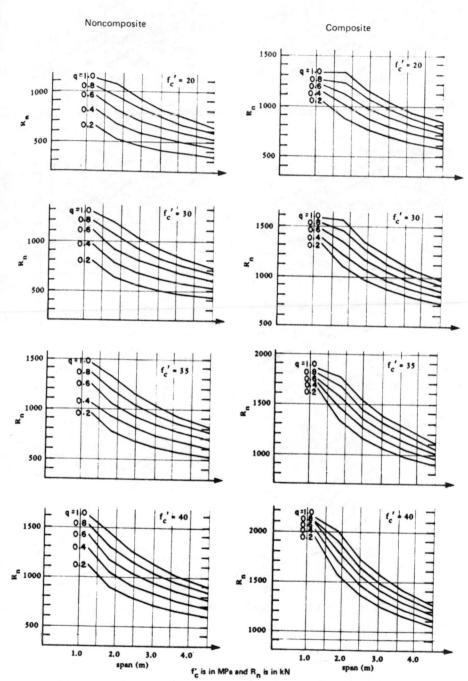

Figure 6.3c Charts for ultimate strengths of 200-mm concrete deck slabs.

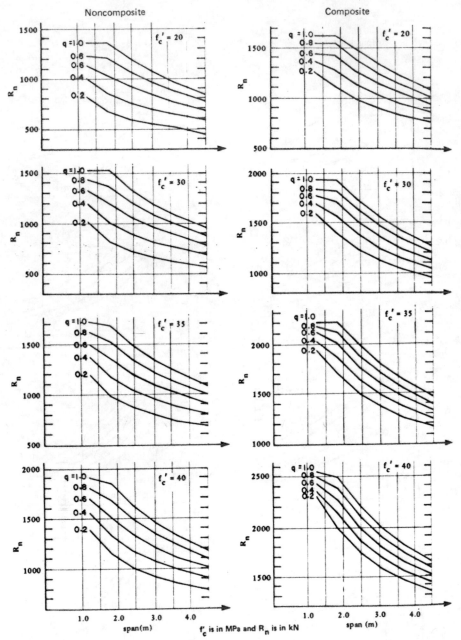

Figure 6.3d Charts for ultimate strengths of 225-mm concrete deck slabs.

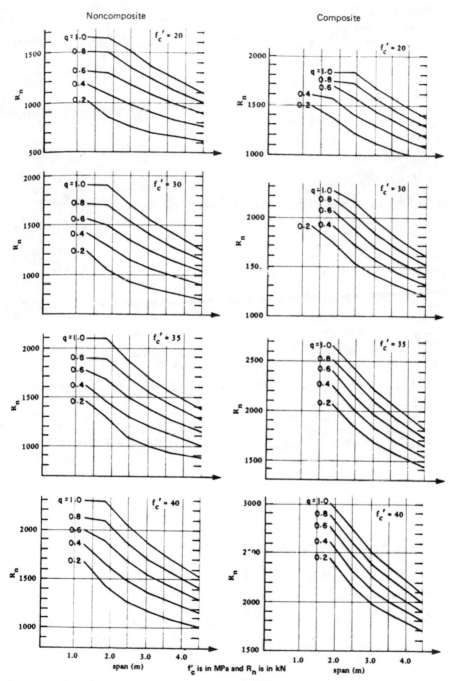

Figure 6.3e Charts for ultimate strengths of 250-mm concrete deck slabs.

2. Calculate the value of θ, using Eq. (2.8), and that of γ from the following expressions:

$$\gamma = \frac{2b - 7.5}{9.0} \quad \text{in metric units} \tag{6.2a}$$

$$\gamma = \frac{2b - 24.6}{29.5} \quad \text{in U.S. Customary units} \tag{6.2b}$$

with a minimum of 0.0 and maximum of 1.0. The bridge width $2b$ is in meters for Eq. (6.2a) and in feet for Eq. (6.2b).

3. Calculate the value of F from the following expression:

$$F = 0.46(\theta)^{0.5}(1 + 0.55\gamma) \tag{6.3}$$

4. Obtain the maximum live-load transverse moment M_{yG} from:

$$M_{yG} = FM_{x,av} \tag{6.4}$$

5. The value of M_{yG} as obtained above is applicable at the bridge centerline for the middle two-thirds of the bridge, as shown in Fig. 6.1. Governing values of live-load transverse moments at other locations can be found through use of Eq. (6.1) and by reference to Fig. 6.1. Alternatively, as specified in the Ontario code, M_{yG} may be used for design purposes over the central one-half of the bridge width and the central two-thirds of the span, as shown in Fig. 6.4. For the remaining portion of the bridge, a value of $0.75\,M_{yG}$ should be used.

Example 6.1 A slab bridge with a width of 32 ft (9.75 m) and span of 40 ft (12.19 m) is analyzed for the AASHTO HS 20 vehicle. All calculations for this example are given in U.S. Customary units.

The maximum moment due to one lane of loading for a 40-ft span is 449,800 lb · ft. Hence, the average longitudinal moment is 449,800/32 = 14,060 lb · ft/ft.

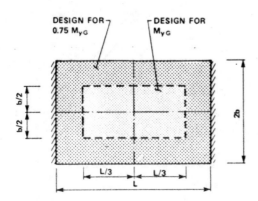

Figure 6.4 Transverse moment zones for design.

From Eq. (2.8), θ is given by

$$\theta = \frac{32}{2(40)} = 0.40$$

From Eq. (6.2), γ is given by

$$\gamma = \frac{32 - 24.6}{29.5} = 0.25$$

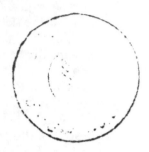

From Eq. (6.3),

$$F = 0.46(0.4)^{0.5}(1 + 0.55 \times 0.25) = 0.33$$

From Eq. (6.4) the maximum live-load transverse moment M_{yG} is given by:

$$M_{yG} = 0.33(14,060) = 4640 \text{ lb} \cdot \text{ft/ft}$$

With M_{yG} known, the transverse moment M_{yg} anywhere across the bridge width can be calculated by Eq. (6.1).

6.4 Analytical Methods for Slab-on-Girder Bridges

Methods of analysis given in this section were primarily developed by Kryzevicius [6]; their developmental background is reported in Ref. 7. The methods are applicable only to those bridges which can be realistically idealized as orthotropic plates. Hence, a bridge that is analyzed by the methods given here should conform to the limitations listed in Sec. 4.1.

As discussed in Sec. 3.4, the live-load transverse moments in slab-on-girder bridges are assumed to be the sum of global and local transverse moments. Methods are given below for obtaining these two components of the total transverse moment.

Background for Local Moments

Influence surfaces for plates of constant thickness and infinite width, given in Ref. 9, were used to calculate moments due to wheel loads.

It is well known that moment intensities under concentrated loads increase rapidly in size as the contact area of the load is reduced. Hence realistic estimates of moment intensities under a concentrated load can only be obtained when the contact area of the concentrated load is closely correct. The heaviest wheels of both the AASHTO and Ontario design vehicles have the same contact area, namely, 600×250 mm (24×10 in). The effective loaded area at the middle surface of a concrete slab 200 mm (8 in) thick is obtained by applying a dispersion angle of 45° from the boundaries of the contact area at the top surface through

75 mm (3 in) of wearing surface and 100 mm (4 in) of slab thickness, as shown in Fig. 6.5. The result is an effective loaded area at the middle surface of the slab in the form of a rectangle 950 × 600 mm (37 × 24 in).

Figure 6.6 shows the assumed boundary conditions for local transverse moment analysis in exterior and interior deck panels. These simple boundary conditions of full rotational fixity are justifiable because symmetrical loads on the two sides of an interior girder can be taken to cause zero rotation above the girder.

The influence surfaces used for analysis are, in fact, independent of the span of the panel. However, when calculating moments due to concentrated loads of finite contact area, the span of the panel does come into consideration in relation to the dimensions of the contact area. Plates having spans from 2.25 to 5.00 m (7.4 to 15.4 ft), at intervals of 0.25 m (0.82 ft), were analyzed for maximum moments due to a unit load having a contact area of 950 × 600 mm (37 × 24 in), using the influence surfaces. The method is semigraphical, and a planimeter was used to calculate areas enclosed by the moment coefficient contours. Integrated moments under the unit load were than obtained by calculating the total volume under the moment surface.

Maximum values of both negative and positive local moments were calculated separately for AASHTO and Ontario loadings, for each of the panel spans.

Local Moments, AASHTO Loading

The maximum values of local transverse moments M_{yL} due to H 20 and HS 20 loadings can be read directly from Fig. 6.7, for any given spacing S of the girders. The curves in the figure, besides providing the values of M_{yL}, also provide values of local longitudinal moment M_{xL}. It can be seen that the local slab moments in the longitudinal direction are not small (as they are usually believed to be), and should not be ignored. Attention is

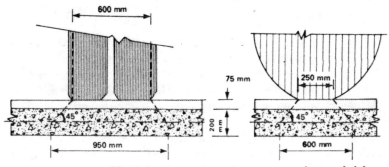

Figure 6.5   Assumed load dispersion through wearing surface and slab.

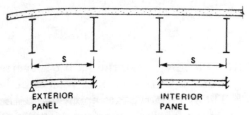

Figure 6.6 Assumed boundary conditions for local bending analysis.

drawn to the fact that the negative local moments are much larger than their positive counterparts. It is noted that the M_{yL} values given in Fig. 6.7 are unfactored.

Local moments due to H 15 and HS 15 loadings can be obtained by multiplying the corresponding moments due to H 20 loading by a factor of 0.75. A multiplier of 0.5 is similarly required to obtain the moments for H 10 and HS 10 loadings.

Figure 6.7 provides local transverse moment values in kilonewton-

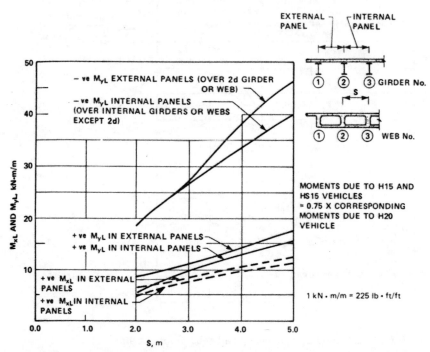

Figure 6.7 Local transverse moments and related longitudinal moments due to H 20 and HS 20 loading.

meters per meter (kN · m/m). These values can be converted into pound-feet per foot (lb · ft/ft) by multiplying them by 225.

Local Moments, Ontario Loading

In a similar manner to the AASHTO loading case, the local transverse moments due to Ontario loading shown in Fig. 4.6 can be read directly from the curves given in Fig. 6.8. The same curves are applicable to the design truck and all three levels of the evaluation vehicle. This is so because the local moments are mainly governed by the two closely spaced heavy axles of the vehicles, and these axles are common to all vehicles. It is noted that the values of M_{yL} given in Fig. 6.8 are unfactored.

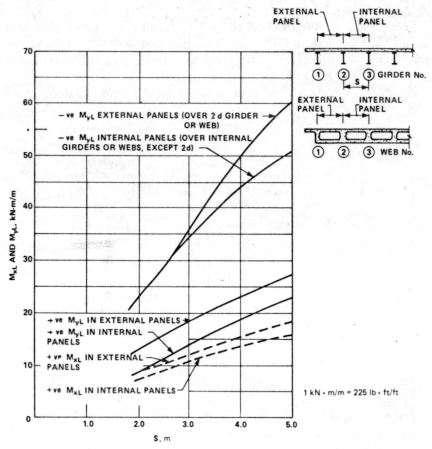

Figure 6.8 Local transverse moments and related longitudinal moments due to OHBDC truck loads.

Local Moments in Box Girders

It is customary to obtain local transverse moments in the top concrete flanges of box girders with the help of influence surfaces, using the same boundary conditions as are shown in Fig. 6.6. The curves of Figs. 6.7 and 6.8 can, therefore, also be used for obtaining local transverse moments in box girder bridges.

It should be noted that there does not exist at present a reliable simplified method for obtaining global transverse moments in box girder bridges. A suitable refined method should be used for obtaining these global moments.

Background for Global Moments

Most slab-on-girder bridges develop only sagging live-load transverse global moments. Negative live-load global moments of any consequence can only develop in very wide bridges, usually those having four or more lanes. These negative moments, which can be of the same order of magnitude as the positive global moments, are caused by loading the bridge with two vehicles, each one placed as close as possible to the two longitudinal free edges. The maximum negative global moment, like its positive counterpart, occurs at or near the longitudinal centerline of the bridge. A method for determining the maximum negative moment is not given here. The recommended procedure is to design the deck slab of a very wide bridge for the same magnitude of negative global transverse moment as the positive one. Local negative moments should not be added to negative global moments.

Global Moments, AASHTO Loading

The following steps are required for calculating global transverse moments due to AASHTO loading for slab-on-girder bridges:

1. Calculate θ according to Eq. (2.3) for noncomposite bridges and Eq. (2.4) for composite ones.

2. Corresponding to the value of θ and the span and width of the bridge, read from Fig. 6.9 the value of global transverse moment intensities, M_{yG}, for HS 20 loading. Use linear interpolation for values of width and θ which are different from those for which the curves are given. For a loading other than HS 20 loading, multiply the value of transverse moment intensity obtained above by M/M_{HS20}. In this expression, M is the maximum moment for the span under consideration due to one lane of the loading concerned and M_{HS20} is the corresponding maximum moment

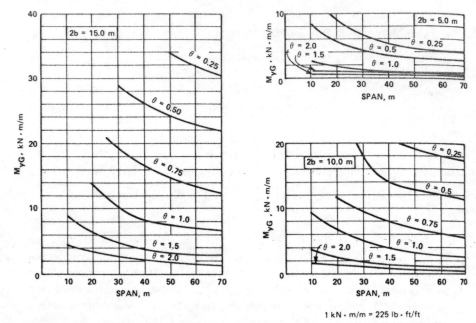

Figure 6.9 Global transverse moments in slab-on-girder bridges due to HS 20 loading.

due to one lane of HS 20 loading. Figure 4.17 will be found useful in obtaining the factor M/M_{HS20}.

The global moment intensities obtained above are unfactored.

Global Moments, Ontario Loading

The steps in the calculation of global transverse moments due to Ontario loading are the same as for AASHTO loading, except that Fig. 6.10 should be used for moments due to the design vehicle and the level 3 evaluation vehicle. For obtaining the moments due to level 2 and level 1 evaluation vehicles, the moments obtained from Fig. 6.10 should be multiplied by the ratio of maximum moments due to the evaluation vehicle under consideration and to the design vehicle.

Example 6.2 Transverse moments due to H 15 loading are calculated below for the same bridge that was analyzed in Example 4.2, the cross section of which is shown in Fig. 6.11.

The girder spacing of the bridge is 8.25 ft, or 2.5 m. Corresponding to this spacing the negative local moment intensity M_{yL} due to HS 20 loading is found from Fig. 6.7 to be 23 kN · m/m, which is 5175 lb · ft/ft in USCS units, for both external and internal panels. The corresponding value for HS 15 loading is three-quarters of this, that is, 17.25 kN · m/m, or 3881 lb · ft/ft. Similarly, the positive local moment intensities M_{yL} due to HS 20 loading are found to be 10 kN · m/m (2250 lb · ft/ft) and 7.5 kN · m/m (1800 lb · ft/ft) for external and internal panels, respectively; the corresponding values for HS 15 loading are three-quarters of these, that is, 7.5 and 6 kN · m/m (1690 and 1350 lb · ft/ft), respectively.

From Example 4.2, θ for this bridge is 0.87. The span and width of the bridge are 85.0 and 40.7 ft, or 25.9 and 12.4 m, respectively. For a span of 25.9 m and θ of 0.87, Fig. 6.9 gives global moment intensities M_{yG}, for HS 20 loading, of 16.7 and 8.1 kN · m/m for bridge widths of 15.0 and 10.0 m, respectively. Hence the corresponding value for HS 20 loading for a bridge width of 12.4 m is given by interpolation as

$$8.1 + \left(\frac{12.40 - 10.0}{15.0 - 10.0}\right)(16.7 - 8.1) = 12.2 \text{ kN} \cdot \text{m/m}$$

From Fig. 4.17, the maximum moments in a span of 85 ft, due to one-half the HS 15 and HS 20 loadings, respectively, are approximately 370 and 630

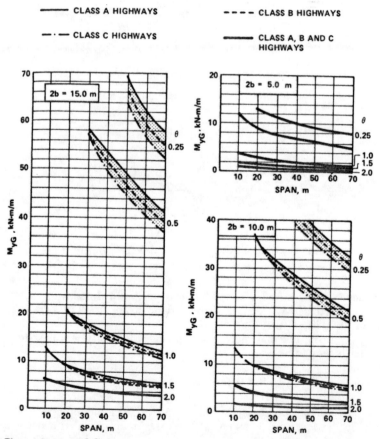

Figure 6.10 Global transverse moments in slab-on-girder bridges due to Ontario loading.

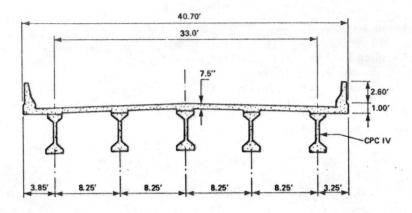

	--3.88		-3.88		-3.88			− ve LOCAL MOMENT 1000 lb · ft/ft
	1.69		1.35		1.35		1.69	+ ve LOCAL MOMENT 1000 lb · ft/ft
0.0	0.62	1.15	1.50	1.62	1.50	1.15	0.62	0.0 + ve GLOBAL MOMENT 1000 lb · ft/ft
0.0	2.31	-2.73	2.82	-2.26	2.82	-2.73	2.31	0.0 TOTAL TRANSVERSE MOMENTS, 1000 lb · ft/ft

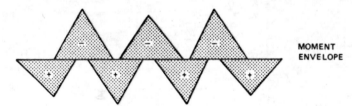

MOMENT
ENVELOPE

Figure 6.11 Transverse moments applicable for the middle two-thirds of the span.

kip · ft. Hence M/M_{HS20} is $370/630 = 0.59$. The maximum global transverse moment M_{yG} due to HS 15 loading is thus found to be $0.59(12.2) = 7.2$ kN · m/m, or 1620 lb · ft/ft.

By using Eq. (6.1), the maximum global moment intensity at a next-to-outside girder is seen to be

$$1620 \cos \frac{8.25\pi}{33.0} = 1145 \text{ lb} \cdot \text{ft/ft}$$

Using the same formula [(Eq. (6.1)] the global transverse moment intensity at the external girder locations is taken as zero. Maximum global transverse moment intensities at other locations are calculated by using Eq. (6.1) and are shown in Fig. 6.11, which also shows the total transverse moment intensity at various locations and the moment envelope. It should be noted that these transverse moments are unfactored, and in particular have not been modified to take account of the impact factor.

References

1. American Association of State Highway and Transportation officials (AASHTO): *Standard Specifications for Highway Bridges,* Washington, D.C., 1977.
2. Batchelor, B. de V., Hewitt, B. E., Csagoly, P. F., and Holowka, M.: *An Investigation of the Ultimate Strength of Deck Slabs of Composite Steel/Concrete Bridges,* TRR 664, Transportation Research Board, Washington, D.C., 1978, pp. 162–170.
3. Csagoly, P. F., Holowka, M., and Dorton, R.: *The True Behavior of Thin Concrete Slabs,* TRR 664, Transportation Research Board, Washington, D.C., 1978, pp. 171–179.
4. Cusens, A. R., and Pama, R. P.: *Bridge Deck Analysis,* Wiley, London, 1975.
5. Jaeger, L. G., and Bakht, B.: The grillage analogy in bridge analysis, *Canadian Journal of Civil Engineering,* 9(2), 1982, pp. 224–235.
6. Kryzevicius, S.: Private communication, 1982.
7. Kryzevicius, S., and Bakht, B.: *Transverse Moments in Slab-on-Girder Bridges,* Structural Research Report SRR-83-5, Ministry of Transportation and Communications, Downsview, Ontario, Canada, 1983.
8. Ministry of Transportation and Communications: *Ontario Highway Bridge Design Code (OHBDC),* 2d ed., Downsview, Ontario, Canada, 1983.
9. Pucher, A.: *Influence Surfaces of Elastic Plates,* Springer-Verlag, New York, 1964.

7

METHODS FOR TRANSVERSE SHEAR

7.1 Introduction

The response known as transverse shear, V_y, is illustrated in Fig. 3.14, and is usually of significance only in those bridges in which it is the predominant means of load transference between one longitudinal beam and another. The methods given in this chapter, strictly speaking, apply only to such bridges. If the methods are used for other bridges, in which transverse flexure is mainly responsible for transverse load distribution, then the values of transverse shear that are predicted will always be on the high side. The extent to which transverse shear is overestimated will depend upon the transverse flexural stiffness of the cross section as a whole.

It should further be noted that for the methods given in this chapter to be applicable, the transverse spacing between adjacent longitudinal beams, across which the transverse shear V_y operates, should be relatively small, as is the case, for example, in multibeam bridges (Fig. 2.18). This restriction is imposed because the method provides maximum V_y values in the immediate vicinity of concentrated loads. If the spacing between adjacent longitudinal beams is large, as it is, for example, in multispine bridges, then the peak value of V_y rapidly diminishes as one moves transversely away from the concentrated load position. In such cases the values calculated by the simplified methods give a highly conservative picture of the longitudinal distribution of V_y.

The methods given here were developed, as reported in Ref. 2, using the articulated plate theory of Ref. 4. It was found that the maximum intensity of V_y, besides depending on the parameter β defined in Sec. 1.6, depends also on the span and width of the bridge.

7.2   Method for AASHTO Loading

The following steps are required in obtaining the unfactored design values of live-load transverse shear intensity in multibeam bridges.

1. Calculate the value of β, using Eq. (2.18), according to the procedure given in Sec. 2.3.

2. Corresponding to the value of β and the span and width of the bridge, obtain the value of maximum transverse shear intensity for H 20 and HS 20 vehicles from Fig. 7.1. Use linear interpolation for values of β and width $2b$ that are different from those for which the curves are given.

3. For H 15 and HS 15 vehicles, multiply by 0.75 the value of maximum V_y obtained in step 2 for H 20 and HS 20 vehicles. For H 10 and HS 10 vehicles use a multiplier of 0.5.

Example 7.1 Maximum intensity of live-load transverse shear due to the HS 20 vehicle is calculated for a 60-ft (183-m) span multibeam bridge, the cross section of which is shown in Fig. 7.2. Using Eqs. (2.20) and (2.21) the values of D_x and D_{xy} for the bridge are found to be $902E$ and $2101G$, respectively, in inches.

Taking Poisson's ratio for concrete to be approximately 0.15, the value of E_c/G_c is 2.3, and hence β is obtained from

$$\beta = \pi \left(\frac{31.5 \times 12}{60.0 \times 12} \right) \left(\frac{902 \times 2.3}{2101} \right)^{0.5} = 1.64$$

For β equal to 1.64 and L equal to 60 ft, the governing values of V_y are obtained from Fig. 7.1 for bridge widths of 25 and 33 ft, respectively; these values are 170 and 210 lb/in, respectively. The governing value for the 31.5-ft-wide bridge is obtained by linear interpolation as follows:

$$V_y = 170 + \left(\frac{31.5 - 25.0}{33.0 - 25.0} \right) (210 - 170) = 202 \text{ lb/in}$$

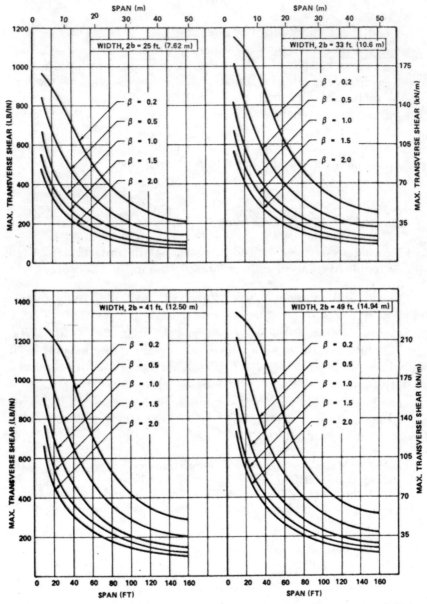

Figure 7.1 Maximum shear intensity in multibeam bridges due to H 20 and HS 20 vehicles.

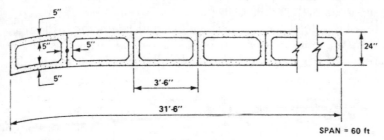

Figure 7.2 Part section of a multibeam bridge.

7.3 Method for Ontario Loading

The method for Ontario loading is the same as for AASHTO loading, with the unfactored governing values of V_y being read from Fig. 7.3 instead of Fig. 7.1. The governing values of V_y given in Fig. 7.3 are applicable both for the design vehicle and for all three levels of the evaluation vehicle.

7.4 Some Observations

It is noted that the value of V_y, obtained from Fig. 7.1 or 7.3, is the maximum value that occurs anywhere on the bridge. This absolute maximum is obtained when the vehicle is as eccentrically positioned as possible, and occurs near the outer line of wheels. The maximum V_y at the bridge centerline can be smaller by up to 25 percent in bridges having three or more lanes.

The maximum V_y value corresponds to the single heaviest axle of the design vehicle, and is proportional to this heaviest axle load. The curves given in Figs. 7.1 and 7.3 can, therefore, also be utilized for vehicles other than AASHTO and Ontario vehicles, provided that the values of V_y obtained from the figures are multiplied by the ratio of the heaviest axle load of the vehicle under consideration to that for which the curve was drawn.

Nail-Laminated Bridges

Bridges composed of longitudinal wood laminates which are nailed together can be realistically idealized as articulated plates during the early stages of their lives. After such bridges have been subjected to vehicular traffic for 2 or 3 years, the holes which contain the nails enlarge, thus permitting a given laminate to deflect a little on its own before influencing the adjacent ones. Such "lost motion" movements of individual laminates make it impossible to idealize the structure by a simple continuous medium such as an orthotropic plate.

Estimates of live-load transverse shear in a nail-laminated bridge can be obtained by the methods given in Secs. 7.2 and 7.3, using Eqs. (2.24) for calculation of the plate parameter. Timber bridges, being torsionally

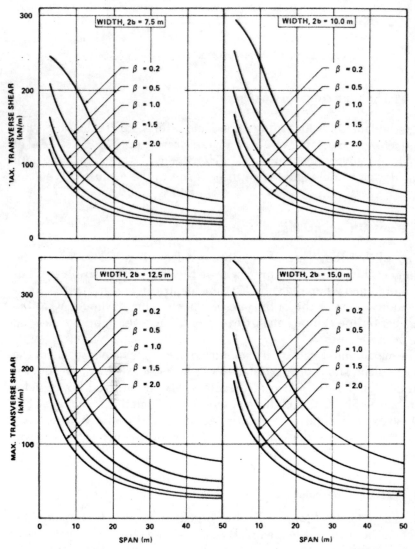

Figure 7.3 Maximum transverse shear intensity in multibeam bridges due to OHBDC truck.

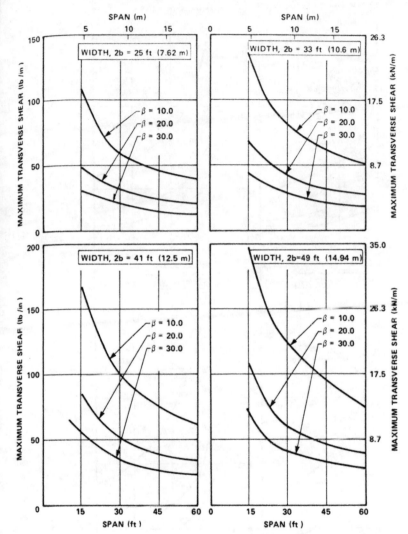

Figure 7.4 Transverse shear intensity in torsionally weak articulated plates due to AASHTO H 20 and HS 20 vehicles.

weaker, have much larger values of β than multibeam bridges. These β values fall outside the range of values given for the charts in Figs. 7.1 and 7.3. For larger values of β, the corresponding charts are given in Figs. 7.4 and 7.5. These charts should be used for obtaining V_y in timber bridges. The values of V_y obtained in this manner are valid only when the bridge is relatively new. With increasing usage of the bridge, with concomitant enlargement of the nail holes, the actual maximum intensity of transverse shear will decrease.

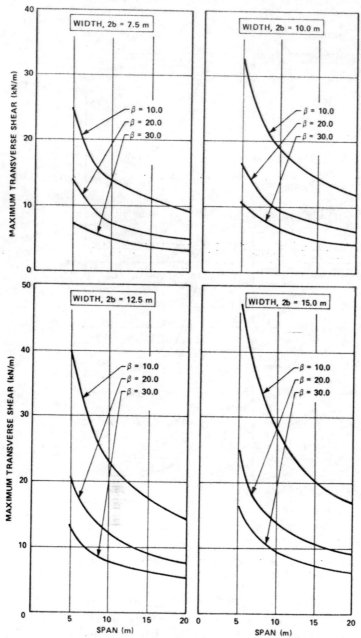

Figure 7.5 Transverse shear intensity in torsionally weak articulated plates due to Ontario design truck.

Transversely Prestressed Wood Bridges

In bridges composed of longitudinal wood laminates which are transversely post-tensioned (Fig. 2.14), it is necessary to calculate live-load interlaminate shear as one of the design responses. This response is the same as transverse shear V_y, and hence can be obtained by calculating β from plate rigidities calculated by Eq. (2.25) and then using Fig. 7.4 or Fig. 7.5 as needed.

The idealization of these bridges as articulated plates, i.e., as plates with zero D_y, has been justified by rigorous analysis. It has been shown in Ref. 2 that when D_y of an orthotropic plate is very small as compared with D_x, then the actual value of D_y is of little consequence. In a transversely prestressed laminated-wood bridge the value of D_y is typically about one-twentieth that of D_x, this small value arising from the fact that the effective value of Young's modulus of wood in the transverse direction is only a small fraction of that in the longitudinal direction. When the ratio D_y/D_x is as small as this, the load distribution characteristics of the bridge are virtually the same as those obtained by taking D_y equal to zero. From the design standpoint it is reassuring to note that the small error involved is on the safe side.

References

1. American Association of State Highway and Transportation Officials (AASHTO): *Standard Specifications for Highway Bridges*, Washington, D.C., 1977.
2. Bakht, B., Jaeger, L. G., and Cheung, M. S.: Transverse shear in multi-beam bridges, *Journal of Structural Division*, ASCE, 109(4), 1983, pp. 936–949.
3. Ministry of Transportation and Communications: *Ontario Highway Bridge Design Code (OHBDC)*, 2d ed., Downsview, Ontario, Canada, 1983.
4. Pama, R. P., and Cusens, A. R.: Edge stiffening of multi-beam bridges, *Journal of Structural Division*, ASCE, 93(ST2), 1967, pp. 141–161.

ANALYSIS OF MULTISPAN AND VARIABLE-SECTION BRIDGES

8.1 Introduction

Most of the methods of analysis given in earlier chapters were developed for single-span bridges which conform to the limitations stipulated in Sec. 4.1. As was briefly explained in that section, some of the limitations can be relaxed following an educated assessment of the problem. However, the application of simplified methods to the analysis of continuous-span bridges requires a more extensive treatment, which is given in this chapter.

It is noted that multispan bridges frequently exhibit significant variation in the longitudinal flexural rigidity of the bridge as one moves along the span. For example, the provision of haunches in the neighborhood of internal supports is associated with large local increases in flexural rigidity. Again, steel plate girder bridges, despite near-constant depth of construction, can have markedly varying moments of inertia at different points along the span because of changes in the dimensions of flanges and web plates. The variation in the moment of inertia of an actual three-span plate girder, which is composite with the concrete deck slab, is presented in Fig. 8.1. It can be seen that the moment of inertia of the girder, and

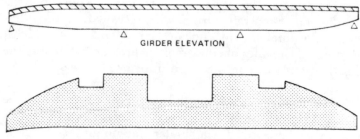

GIRDER ELEVATION

MOMENT OF INERTIA OF THE COMPOSITE GIRDER

Figure 8.1 Nonuniform moment of inertia of a composite plate girder.

hence the total longitudinal flexural rigidity of the bridge, is not even approximately uniform. In such a situation the engineer should not have to rely on unassisted engineering judgment in deciding whether the simplified methods of analysis are applicable to the kind of bridge under consideration. Accordingly, the problem of bridges with varying cross sections is also dealt with in this chapter.

8.2 Mechanics of Behavior of Continuous Bridges

As a preliminary to applying the simplified methods to continuous bridges, it is instructive to examine the mechanics of their behavior. As a vehicle for doing this, a certain slab-on-girder bridge with two spans and three lanes was analyzed by the grillage analogy method [2], and the locations of the points of contraflexure across the bridge width were examined for various load cases. A locus of the points of contraflexure is referred to here as the *line of contraflexure*.

Lines of Contraflexure

Figure 8.2a shows in plan the locations of the lines of contraflexure for different numbers of vehicles. In each case the vehicles are so positioned as to produce the maximum positive moments in a span. It is interesting to note that the line of contraflexure remains reasonably straight and parallel to the support when all the lanes are loaded. However, when fewer lanes are loaded, the line "curls" toward the intermediate support, but only in regions where the intensity of longitudinal moments is expected to be small. In regions where the intensities of longitudinal moments are large, and are therefore of primary interest, the line of contraflexure remains fairly straight and parallel to the support, regardless of the number of loaded lanes.

As can be seen in Fig. 8.2b the same trend is repeated when the bridge is loaded for maximum negative moments over a support, with the difference that the line of contraflexure now curls away from the intermediate support as the number of loaded lanes is reduced. It is found that the straight portions of the lines of contraflexure are at substantially the same places as the points of contraflexure in beams of constant cross section.

The analysis described above was carried out for the case in which the longitudinal flexural rigidity of the bridge was uniform along its length. The analysis was then repeated for a similar slab-on-girder bridge in which the girder moment of inertia was assumed to be constant from the simple support to the midspan, and then to increase parabolically to three times the uniform value at the intermediate support. It was found

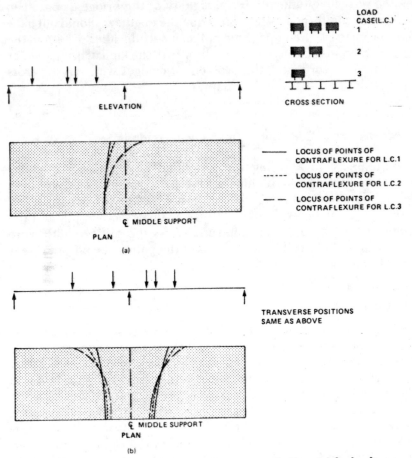

Figure 8.2 Points of contraflexure in a continuous bridge: (a) for loading on one span; (b) for loading on two adjacent spans.

that all lines of contraflexure for this bridge were closer to being straight than the corresponding lines for the bridge with uniform flexural rigidity.

A significant premise, which was confirmed by further analysis, can be drawn from the studies just described: For simplified analysis of the transverse distribution of longitudinal moments, each of the relevant lines of contraflexure can be assumed to be replaced by a straight simple support parallel to the actual support. Thus, the equivalent spans for the various cases can be obtained by analyzing the bridge as a beam as follows:

1. For positive moments in an end span, the equivalent span is taken to be the distance between the end support and the point of contraflexure corresponding to loading for maximum positive moments in the end span.

2. For negative moments over an internal support, the equivalent span is taken to be the distance between the two points of contraflexure on either side of the support, the two points being those corresponding to the loading for maximum negative moments over the support under consideration.

3. For positive moments in an internal span, the equivalent span is taken to be the distance between the two points of contraflexure in the span corresponding to the loading for maximum positive moments.

These equivalent spans are diagrammatically shown in Fig. 8.3.

It is noted that the equivalent spans are to be used only for obtaining the transverse distribution effects of the various responses, and not as an alternative to continuous-beam analysis. Thus, the total longitudinal moment at any cross section of a continuous bridge should be obtained by treating the bridge as a continuous beam, and the equivalent spans should then be used for obtaining the transverse distribution of this total moment.

The equivalent span basically affects only four characterizing parameters, namely, θ [Eq. (2.2)], β [Eq. (2.18)], δ [Eq. 2.28)], and λ [Eq. (2.34)]. Examination of the various charts in which these parameters are used readily shows that small changes in the values of these parameters have negligible effects on the magnitudes of the relevant responses. Since the length of span directly affects the values of these parameters, it is reasonable to conclude that a small change in the length of the equivalent span will have a negligible effect on the distribution of the responses. Hence, one is led to postulate that it is sufficiently accurate to determine the positions of the points of contraflexure under uniformly distributed loads rather than under various vehicle loadings.

In testing out the postulate, a number of two-span and three-span

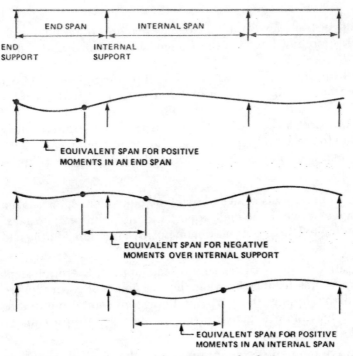

Figure 8.3   Equivalent spans for continuous bridges.

beams with different span lengths were analyzed and equivalent spans were obtained under vehicle loads and corresponding equivalent uniformly distributed loads. It was found that in most cases the equivalent spans under the vehicle loads were smaller than those under the uniformly distributed loads. Nevertheless, the variation was found to be small, being always less than 8 percent. A selection of results for some equispan beams is shown in Fig. 8.4. It can be seen that the degree of error in estimating the equivalent span by the uniformly distributed load method diminishes as the lengths of the spans increase.

An underestimation of span lengths may give answers that are in error on the unsafe side. However, as will be shown subsequently, there are some other factors which more than compensate for this type of error.

The basis of some of the methods given later in this chapter is that, for obtaining the effects of load distribution, each of the negative and positive moment regions of a continuous bridge can be realistically idealized as a rectangular orthotropic plate that is simply supported on two opposite edges. A simply supported edge of the conceptual bridge can either be the actual simple support of the bridge or a line of contraflexure obtained according to the procedure given above. The replacing of a portion of a bridge by the idealized orthotropic plate always leads to conservative (i.e., "safe-side") results, as discussed below.

Load distribution in the negative moment region is considered first. Figure 8.5a shows the deflected shape of this region under loading which induces maximum negative moments over the support. It can be readily visualized that the cross section which experiences the maximum longitudinal moments has nearly zero transverse curvature because of the support constraints, and that the transverse curvature is a maximum at the location of the minimum longitudinal moments, i.e., at the lines of contraflexure. By contrast, the equivalent conceptual bridge (Fig. 8.5b) has the maximum transverse curvature at the cross section with the maximum longitudinal moment; and the transverse section with zero longitudinal moments has zero transverse curvature. It will be appre-

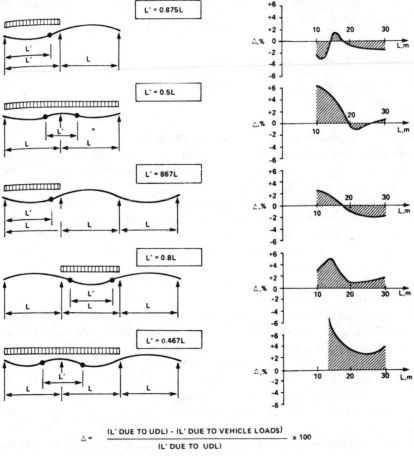

$$\Delta = \frac{(L' \text{ DUE TO UDL}) - (L' \text{ DUE TO VEHICLE LOADS})}{(L' \text{ DUE TO UDL})} \times 100$$

Figure 8.4 Equivalent lengths for multispan beams.

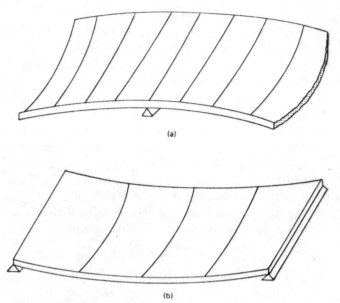

(a)

(b)

Figure 8.5 Differences between the actual negative moment region and the conceptual bridge: (*a*) deflected shape of the negative moment region; (*b*) deflected shape of the equivalent conceptual bridge.

ciated that preventing transverse curvature in the transverse section which experiences the maximum longitudinal moments will help to distribute the longitudinal moments more evenly across the width. From this, it can be concluded that if one replaces the negative moment region of a multispan bridge by a conceptual bridge in which the transverse section with the maximum longitudinal moments is free to take any transverse profile, the result will be a higher-than-actual estimate of the peak intensity of longitudinal moments. It is reassuring to note that the simplification will always lead to safe-side results.

In consideration of the positive moment region contained between the two lines of contraflexure, as shown in Fig. 8.6*a*, for simplified analysis this portion of the actual bridge is replaced by a conceptual bridge in which the lines of contraflexure of the actual bridge are replaced by unyielding supports in the manner shown in Fig. 8.6*b*. The effect of replacing the lines of contraflexure (which do have transverse curvatures) by unyielding supports (which do not) is that the transverse distribution of longitudinal moments becomes "peakier." This phenomenon can be visualized if the conceptual bridge is thought of as being supported on springs instead of rigid supports. In that case, the supported edges have transverse curvatures, as do the lines of contraflexure of the actual bridge, so that this second conceptual bridge, which is shown in Fig. 8.6c, is a closer representation of the actual. From a comparison of load distribution behavior of the two conceptual bridges, it can readily be established that the transverse load distribution in the bridge with spring supports will be more uniform than that in the bridge with line supports. Consequently, it can be confidently expected that the replacing of the actual positive moment region with a conceptual bridge having rigid supports will lead to safe-side results.

Rigorous analysis has indeed confirmed that the simplified methods proposed in Sec. 8.4 always give safe-side results.

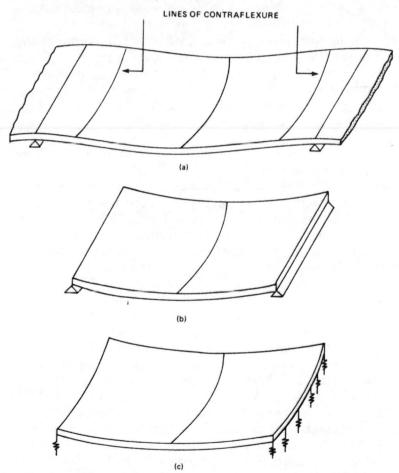

LINES OF CONTRAFLEXURE

(a)

(b)

(c)

Figure 8.6 Differences between the actual positive moment region and conceptual bridges: (a) deflected shape of the positive moment region; (b) deflected shape of the equivalent conceptual bridge on rigid supports; (c) deflected shape of the equivalent conceptual bridge on spring supports.

8.3   Variation of *D* along the Span

The factor D has been the basis of many simplified methods given in this book, and those of the AASHTO and Ontario codes [1, 3]. It is recalled that D has the units of length and is given by the total longitudinal load effect (e.g., total longitudinal moment) due to one line of wheels of the design vehicle divided by the maximum intensity per unit width of the same load effect. The basic assumption behind many of the simplified methods given here is that for a given load effect the value of D remains substantially constant along the span. Yet, as was shown in Sec. 5.1 this is not the case in real life. It is, therefore, instructive to examine the actual variation of D values for longitudinal shear along a two-span continuous bridge, as obtained by the computer-based grillage analogy method, and to investigate what effect their variation has on the final results.

The longitudinal positions of the wheel loads, and the corresponding nodal loads used in the grillage analysis, are shown in Fig. 8.7a. The D values for longitudinal shear corresponding to a certain transverse position of the vehicle were calculated at a different longitudinal position from the grillage analysis results and are shown in Fig. 8.7b. It can be seen that the D values are far from being uniform along the span, varying between 1.5 and 4.1 m (4.92 and 13.45 ft). At first sight this large variation of the D values seems to negate the very basis of the simplified methods. Fortunately this is not the case, for reasons given below.

A D value is usually obtained at a location where the total longitudinal load effect is a maximum. As shown in Fig. 8.7c, the maximum longitudinal shear due to the vehicle is a maximum just to the right of the middle support. The D value corresponding to this location is almost 2.5 m (8.20 ft). Figure 8.7d shows the variation of the longitudinal shear in a girder as obtained by grillage analysis, and its comparison with shear values obtained by using the single D value of 2.5 m (8.20 ft). It can be seen that the large variation of the actual D values results in only minor changes in the shear values obtained by the simplified method, in which no variation of D along the span is accounted for. This is because the large variations of the D values are associated with regions in which the longitudinal shear intensity is very small, so that even large percentage errors in the estimation of the D values in these regions result in small errors in calculated load effects.

It is interesting to note in Fig. 8.7d that the simplified method for the most part predicts slightly larger load effects than are obtained by rigorous analysis, and that where it predicts smaller values, the degree of underestimation is very small. The simplified methods, therefore, can be used with confidence even if the D values vary considerably along the span.

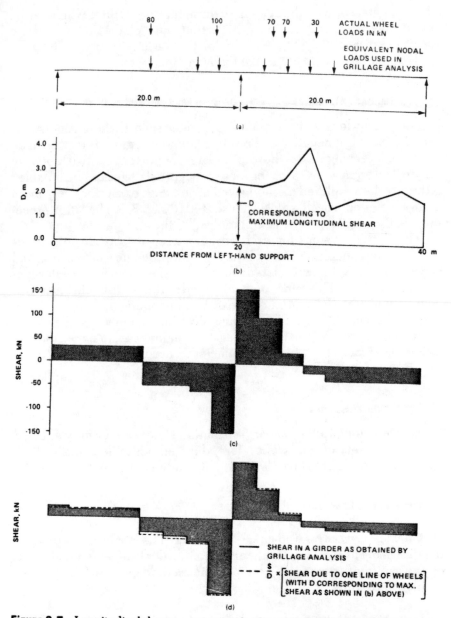

Figure 8.7 Longitudinal shear variation in a bridge. (*a*) Wheel loads on the bridge. (*b*) Shear *D* values along the span. (*c*) Shear due to one line of wheels along the span. (*d*) Comparison of girder shear obtained by grillage analysis and by the simplified method.

8.4   Shallow Superstructures

This section deals with the analysis of those superstructures which can be realistically idealized as "normal" orthotropic plates and which are identified in Sec. 2.2. The analysis for the various responses in these bridges can be carried out in the following manner.

Longitudinal Moments

For obtaining longitudinal moments in multispan shallow superstructures, each of the negative and positive moment regions is treated as an individual rectangular orthotropic plate supported on two opposite edges. The equivalent spans for these conceptual plates can be obtained either by the simplified method given in Sec. 8.2, or alternatively by an even simpler approach which is illustrated in Fig. 8.8. The latter procedure is also approved by the Ontario code [3].

The α and θ values for each of the conceptual orthotropic plates should be obtained according to the relevant procedure of Sec. 2.2, and appropriate D_d values calculated according to the relevant procedures of Secs. 4.2 and 4.3. The values of D_d obtained in this way should be used in conjunction with the continuous beam bending moment diagram due to one line of wheels to obtain the design live-load moments.

It is noted that the θ values for negative regions can, in some cases, be larger than 2.5. In such cases it is sufficiently accurate to assume θ equal to 2.5.

Longitudinal Shears

Fortunately, longitudinal shear distribution is not significantly affected by the continuity of spans. Therefore, the methods of Secs. 5.2 and 5.3 apply to continuous bridges without any modification.

Transverse Moments

Empirical method for slab-on-girder bridges   The empirical method for the design of concrete deck slabs of slab-on-girder bridges is not affected by the continuity of spans. Therefore, the method given in Sec. 6.2 also applies to continuous bridges.

Moments in positive moment regions of slab bridges   The simplified analytical method of Sec. 6.3 can also be applied for determining transverse moments in the regions of continuous slab bridges where the longitudinal moments are positive (i.e., those causing tension in bottom

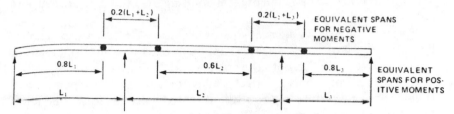

Figure 8.8 Alternative procedure for obtaining equivalent spans.

fibers). For this case the same span should be used as that for the longitudinal moment method. The variation of the maximum transverse moment within the span can be established in the same way as for a single span, i.e., with reference to either Fig. 6.1 or 6.4. In both cases the effective span should be taken as the actual span.

Moments over internal supports of slab bridges Negative live-load transverse moments (i.e., those causing tension in the top fibers) can exist over internal supports of continuous slab bridges even if the internal support is a rigid line support. These negative moments are caused by prevention of the anticlastic curvature which would otherwise occur as a consequence of the longitudinal curvatures, because of the Poisson's ratio of the slab material. It should, however, be noted that the envelope of such transverse moments is approximately of the form shown in Fig. 8.9. For bridges having widths smaller than about 4.0 m, negative transverse moments are negligibly small.

An estimate of the maximum intensity of negative transverse moments over an internal line support can be obtained by multiplying the maximum intensity of the live-load negative moment for internal portions by the Poisson's ratio of concrete (~ 0.15). Moments obtained in this way may be appropriate without modification in the case of closely

CROSS SECTION

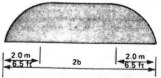

ENVELOPE FOR MAXIMUM NEGATIVE
TRANSVERSE MOMENTS

Figure 8.9 Envelope for maximum negative transverse moments over intermediate line supports of continuous slab bridges.

spaced discrete supports. However, if the supports are widely spaced, account should also be taken of local bending over the discrete supports.

Moments in positive moment regions of slab-on-girder bridges The analytical methods of Sec. 6.4 are also applicable to those multispan slab-on-girder bridges in which both the girders and deck are continuous, as well as to those in which only the deck slab is continuous. It is recommended that for both cases the same equivalent spans be used as were used for the longitudinal moment method (Fig. 8.3 or 8.8).

Moments over internal supports of slab-on-girder bridges The global transverse moments referred to in Sec. 6.1 are caused by differential girder deflections. In regions near the internal supports these differential deflections are either small or completely absent. Hence, the transverse moments over internal supports are closely given by the local moments only, and these can be obtained by the relevant procedure of Sec. 6.4. The lines of contraflexure may be assumed to be the limits of such "local-moment-only" zones.

Example 8.1 A three-span continuous slab bridge is considered for this example. The span lengths and cross section of the two-lane bridge are shown in Fig. 8.10. It is required to calculate live-load longitudinal and transverse moments in the bridge for the Ontario loading corresponding to the ultimate

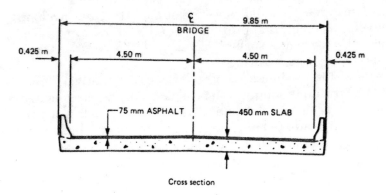

Cross section

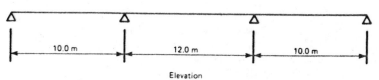

Elevation

Figure 8.10 A slab bridge.

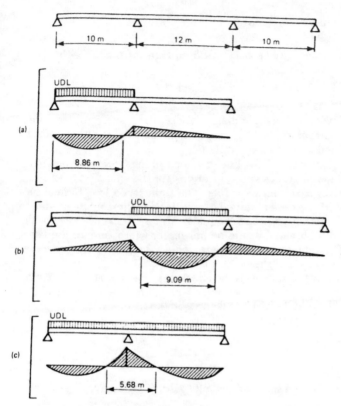

Figure 8.11 Equivalent spans: (*a*) equivalent span for positive moments in an exterior span; (*b*) equivalent span for positive moments in the interior span; (*c*) equivalent span for negative moments over the interior support.

limit state. In accordance with the requirements of the Ontario code, the barrier walls, although monolithic with the slab, are disregarded.

The equivalent spans for various cases of uniformly distributed loads, calculated in accordance with the recommended procedure of Sec. 8.2, are shown in Fig. 8.11. According to Eqs. (2.8),

$$\alpha = 1.0$$

For positive moments in the end span,

$$\theta = \frac{9.85}{2(8.86)} = 0.56$$

For positive moments in the middle span,

$$\theta = \frac{9.85}{2(9.09)} = 0.54$$

For negative moments over an internal support,

$$\theta = \frac{9.85}{2(5.68)} = 0.87$$

The lane width as shown in Fig. 8.10 is 4.5 m. Hence from Eq. (4.6)

$$\mu = \frac{4.5 - 3.3}{0.6} = 2.0 = 1.0 \qquad \text{since } \mu \not> 1.0$$

The D and C_f values corresponding to the above α and θ values, as obtained from the charts of Fig. 4.11b, are given in Table 8.1 together with the D_d values calculated according to Eq. (4.7).

The governing positions of one line of wheels of the Ontario design loading for maximum moments at various locations are shown in Fig. 8.12 along with the respective maximum moments. The unfactored longitudinal moments per unit width are then obtained by multiplying these moments due to one line of wheels by S/D_d, where S is equal to either 1.0 m or 1.0 ft. For the former the moments obtained would be per meter width, and for the latter per foot width. The values of the maximum longitudinal moments per meter width for different cases are shown in Table 8.2.

For positive transverse moments the effective spans are to be those corresponding to positive longitudinal moments. Hence θ for end spans is taken to be 0.56, and for the middle span is taken to be 0.54.

γ is obtained from Eq. (6.2):

$$\gamma = \frac{9.85 - 7.5}{9.0} = 0.26$$

The value of F for the end span, according to Eq. (6.3), is given by

$$F = 0.46(0.56)^{0.5}(1 + 0.55 \times 0.26) = 0.39$$

Similarly, the value of F for the middle span is found to be 0.38.

As shown in Fig. 8.12, the maximum positive moment due to one line of wheels in the exterior span is 258.9 kN $\cdot$ m (190,948 lb $\cdot$ ft). The average moment $M_{x,\text{av}}$ due to one vehicle, i.e., two lines of wheels, is $(2 \times 258.9)/9.85 = 52.6$ kN $\cdot$ m/m (11,824 lb $\cdot$ ft/ft). The maximum live-load positive transverse moment M_{yG} is given by Eq. (6.4):

$$M_{yG} = 0.39(52.6) = 20.5 \text{ kN} \cdot \text{m/m} \ (4608 \text{ lb} \cdot \text{ft/ft})$$

Similarly, the positive M_{yG} for the middle span is found to be 20.3 kN $\cdot$ m/m (4563 lb $\cdot$ ft/ft).

The maximum live-load negative moment in the internal portions over an internal support, as given in Table 8.2, is -117.6 kN $\cdot$ m/m. Assuming that the internal support comprises closely spaced discrete supports, the maximum negative transverse moment over the support is then $-0.15 \times 117.6 = -17.6$ kN $\cdot$ m/m (3956 lb $\cdot$ ft/ft). It is interesting to note that this moment, which is customarily ignored, is in fact about as large as the maximum positive transverse moment in the bridge.

TABLE 8.1 Calculation of D_r Values

Longitudinal reference point	α	θ	C_f, %	External portions		Internal portions	
				D, m	$D_d = D\left(1 + \dfrac{\mu C_f}{100}\right)$, m	D, m	$D_d = D\left(1 + \dfrac{\mu C_f}{100}\right)$, m
Positive moment region in end span	1.0	0.56	16	2.07	2.40	2.02	2.34
Positive moment region in middle span	1.0	0.54	16	2.07	2.40	2.02	2.34
Negative moment region over internal support	1.0	0.87	13	2.05	2.38	1.90	2.20

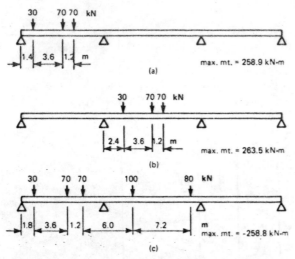

Figure 8.12 Longitudinal positions of the Ontario design vehicle for maximum moments. Load position for (a) maximum positive moment in an exterior span; (b) maximum positive moments in the interior span; (c) maximum negative moment over an interior support.

8.5 Edge Stiffening and Vehicle-Edge Distance

As discussed in Chap. 9, the effects on longitudinal moments of edge stiffening and increased vehicle-edge distance are accounted for by means of a modification factor which depends upon the characterizing parameter λ. This parameter, as defined by Eq. (9.1), involves the length L of the span.

In the case of continuous bridges, λ can be obtained by using the same values of effective spans as those used for longitudinal moments, i.e., those shown in either Fig. 8.3 or 8.8.

8.6 Cellular Bridges and Multibeam Bridges

The effect of cell distortion in cellular bridges is accounted for in the analysis by a modification factor which depends upon θ [Eq. (2.12)] and δ [Eq. (2.28)]. Both these parameters involve the span length. For continuous bridges the same values of effective spans as those used for longitudinal moments (Fig. 8.3 or 8.8) should be used in calculating the θ and δ values.

TABLE 8.2 Unfactored Intensities of Live-Load Longitudinal Moments

Longitudinal reference point	Moment due to one line of wheels, kN·m	External portions		Internal portions	
		D_d, m	Unfactored moment intensity, kN·m/m	D_d, m	Unfactored moment intensity, kN·m/m
Location for maximum positive moment in end span	258.9	2.40	107.9	2.34	110.6
Location for maximum positive moment in middle span	263.5	2.40	109.8	2.34	112.6
Over an internal support	−258.8	2.38	−108.7	2.20	−117.6

8.7   Variable-Section Bridges

Slab-on-Girder Bridges

It was noted in Sec. 8.1 that many multispan slab-on-girder bridges incorporate girders with varying moments of inertia. The effect of this variation on the load distribution characteristics of a bridge can be conveniently studied by carrying out rigorous analysis on specific bridges.

For this study, various two-span slab-on-girder bridges, each having three lanes, were analyzed by the grillage analogy method described in Ref. 2. In a first series of analyses, three bridges having uniform longitudinal flexural rigidities and θ values of 0.5, 1.0, and 1.5 for their respective positive moment regions were analyzed for various vehicle positions. In the second and third series the same bridges were then analyzed twice more each by varying the girder moments of inertia according to the pattern shown in Fig. 8.13. For the second and third series, I_2/I_1 was kept equal to 2.0 and 3.0, respectively, where, as shown in Fig. 8.13, I_1 is the girder moment of inertia for the uniform portion and I_2 is the moment of inertia at the intermediate support.

From the results of the analyses, D values for longitudinal moments, being the total longitudinal moment due to one line of wheels divided by the maximum intensity of longitudinal moments per unit width, were calculated for the nine bridges for both positive and negative moments. The results are given in Table 8.3. Two conclusions can be drawn from the study:

1. The values of D for longitudinal moment decrease only slightly with increase of I_2/I_1.

2. The change of D due to an increase of I_2/I_1 diminishes as the number of vehicles increases.

Other studies have shown that even for bridges in which the girder moment of inertia varies in the manner shown in Fig. 8.1, it is sufficiently accurate, for the analysis of positive moment regions, to assume a mean and uniform value of the longitudinal flexural rigidity.

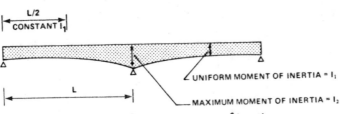

Figure 8.13 A girder of varying moment of inertia.

TABLE 8.3 Effect of Change of Longitudinal Flexural Rigidity in Two-Span Bridges on D Values for Longitudinal Moments

Bridge group no.	No. of vehicles on the bridge	$\dfrac{I_2^*}{I_1}$					
		D, m, over intermediate support			D, m, at midspan		
		1.0	2.0	3.0	1.0	2.0	3.0
1 ($\theta = 0.5$)	1	2.98	2.94	2.92	3.26	3.21	3.19
	2	1.87	1.85	1.85	2.15	2.14	2.14
	3	1.71	1.71	1.71	1.93	1.92	1.92
2 ($\theta = 1.0$)	1	3.74	3.73	3.73	4.14	4.03	3.97
	2	2.22	2.20	2.20	2.47	2.43	2.41
	3	1.86	1.84	1.84	2.00	1.99	1.99
3 ($\theta = 1.5$)	1	4.23	4.21	4.19	4.73	4.57	4.50
	2	2.42	2.42	2.40	2.67	2.61	2.59
	3	1.93	1.93	1.92	2.05	2.03	2.03

* I_2 = moment of inertia of girder over intermediate support; I_1 = moment of inertia of girder in the uniform portion.
Note: $\alpha = 0.15$ for the uniform moment of inertia portion.

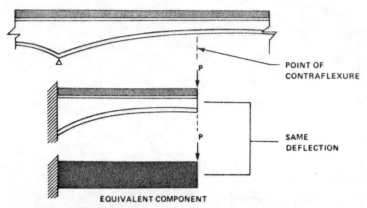

Figure 8.14 Idealization of girder of variable moment of inertia.

For the negative moment regions, however, it is prudent to calculate an equivalent constant longitudinal flexural rigidity by applying the condition shown in Fig. 8.14. The portion of the bridge between the internal support and the point of contraflexure is treated as a cantilever beam carrying a concentrated load at the tip. The equivalent constant flexural rigidity is then that which will give the same tip deflection as is given by the cantilever with varying flexural rigidity. It is noted that small errors in the location of the point of contraflexure have little effect on the final results. Hence, in practice, it is sufficiently accurate to utilize for all cases the same points of contraflexure as are shown in Fig. 8.8.

Slab Bridges

The haunching of the slab in continuous bridges increases the rigidities in both the longitudinal and transverse directions. The increase of the longitudinal flexural rigidity, taken alone, has the effect of making the transverse load distribution more localized. On the other hand, the increase of transverse flexural rigidity helps to distribute the load more evenly across the bridge. The haunching of the slab thus results in two opposing effects which are largely self-canceling in most cases. Thus, in continuous slab bridges it is sufficiently accurate to ignore variation of slab thickness in the longitudinal direction in the calculation of θ.

8.8   Rigid Frame Bridges

A rigid frame bridge, as shown in Fig. 8.15, incorporates abutments which are integral with the floor system. The floor system may be of solid

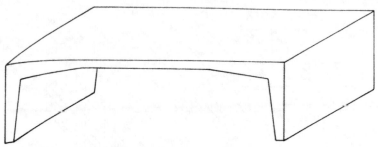

Figure 8.15 A rigid frame bridge.

slab construction, or may comprise a slab that is monolithic with girders. Quite often the slabs in such bridges are of varying depth.

The problem of analyzing the load distribution in a rigid frame bridge can be considerably simplified if the bridge is treated as a continuous bridge, as shown in Fig. 8.16. The equivalent continuous bridge obtained by opening up the "legs" of the rigid frame is different from other continuous bridges in that its end spans are not directly subjected to live loads. The load effects in the end spans due to live loads that are applied to the middle span should understandably be distributed more uniformly than if the live loads were directly applied to the end span itself.

An analysis of a typical solid slab rigid frame bridge by the three-dimensional space frame analogy has shown that the D values for longitudinal moments in the legs of the bridge are not substantially different from those at the "knees." The lines of contraflexure and the D values as obtained from this analysis are shown in Fig. 8.17.

It can be concluded that idealization of a rigid frame bridge as a multispan bridge will not lead to significant errors.

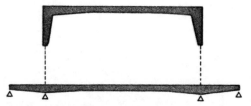

Figure 8.16 Idealization of rigid frame bridge as a continuous bridge.

8.9 Precast Box Culverts

Precast box culverts, because of transportation considerations, are often manufactured in lengths of about 6 ft, or 2 m. A segment of a precast box

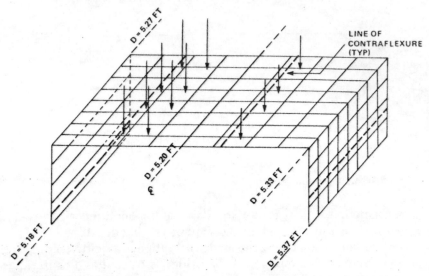

Figure 8.17 *D* values for moments in a rigid frame bridge.

culvert is shown in Fig. 8.18. Spans of such culverts often range from 4.0 to 12.0 ft (1.2 to 3.7 m). The live-load analysis of box culverts is seemingly made difficult by the earth fill on them, and by the fact that the adjacent segments are joined together through a kind of shear connection. Such connections cannot be relied upon to provide continuity of flexural rigidity across the segments; at best they can provide efficient

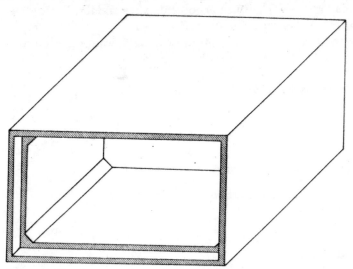

Figure 8.18 A precast box culvert segment.

shear connections. If the widths of the segments are small compared with their spans, then the top portions of the culverts, which carry the live load, can be analyzed as multibeam or shear-connected beam bridges. Segments of larger widths, however, defy such analysis.

If the load distribution behavior of a typical segment is estimated on the basis that no load transfer takes place between the segments and its neighbors, a simple and safe solution can be readily obtained.

Example 8.2 Figure 8.19 shows the cross section of a 6-ft- (1.83-m-) wide segment with a 1.0-ft (0.3-m) earth fill on the top, this depth of fill being normal for such a culvert. The culvert segment shown can accommodate only one line of wheels, it being noted that the distance between the outer extremities of the tire prints of the two lines of wheels is 8 ft 2 in (2.50 m). In both the AASHTO and OHBDC design vehicles, the contact area for one-half of the heaviest axle is 24 × 10 in (600 × 250 mm). The effective loaded area at the middle surface of the culvert slab is usually estimated by assuming that the load disperses at an angle of 45° through the earth fill and the top half of the concrete slab. As shown in Fig. 8.19, the width of this equivalent load is then 56 in (1.42 m). Such a width constitutes about 78 percent of the width of the segment. Rigorous analysis readily confirms that the effect of such a wide load is almost identical with that of a load uniformly distributed across the entire width of the segment.

Shear connections between adjacent segments will distribute loads away from the internally loaded one to adjacent ones; they will, however, also result in attracting the load from the adjacent segments if the latter are also loaded. Owing to constraints on the transverse spacings between vehicles, it is not possible to have more than two consecutive loaded segments, unless the length of each segment is much larger than 6 ft (1.83 m). Therefore, in real-life cases, a shear connection between the segments will result in transferring loads from a loaded segment to segments on both sides; the load attracted by the segment under consideration will come from one side only.

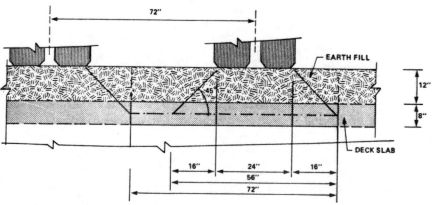

Figure 8.19 Wheel load on a culvert segment.

On balance, a segment would carry more loads if the connections were absent than if they were present.

A safe-side conclusion is to assume D equal to the smaller of 6 ft (1.85 m) and the length of the precast segment. This D value is applicable to both longitudinal moments and longitudinal shears. A uniform D value for the four walls of the box culvert segment does indeed reflect the true behavior of the structure in which most of the live load applied to a segment is sustained by that segment alone in the absence of an effective connection between adjacent segments.

After establishing the value of D, a unit width of the box culvert can be analyzed as a two-dimensional box frame subjected to one line of wheels of the relevant design vehicle multiplied by $1/D$.

References

1. American Association of State Highway and Transportation Officials (AASHTO): *Standard Specifications for Highway Bridges*, Washington, D.C., 1977.
2. Jaeger, L. G., and Bakht, B.: The grillage analogy in bridge analysis, *Canadian Journal of Civil Engineering*, 9(2), 1982, pp. 224–235.
3. Ministry of Transportation and Communications: *Ontario Highway Bridge Design Code (OHBDC)*, 2d ed., Downsview, Ontario, Canada, 1983.

ANALYSIS FOR
EDGE STIFFENING AND
VEHICLE-EDGE DISTANCE

9.1 Introduction

Examples of edge stiffening in slab and voided slab bridges are shown in Fig. 2.28, and those for slab-on-girder and cellular structures in Fig. 2.29. The term *vehicle-edge distance* is illustrated in Fig. 2.27. These two factors, i.e., edge stiffening and vehicle-edge distance, have been eliminated from consideration in the simplified methods of Chap. 4 by assuming that the former is absent from the design and that the latter has a fixed value of about 1.0 m (3.28 ft).

As has been shown in Sec. 3.2, the maximum intensity of longitudinal moments decreases rapidly as the vehicle-edge distance increases. Similarly, appendages near the longitudinal free edges of the bridge, i.e., edge stiffening, by attracting loads to themselves, tend to reduce the intensity of longitudinal moments in the unstiffened portions of the cross section of the bridge. It is noted that the reduction of longitudinal moment intensity which results from an increase in vehicle-edge distance is more pronounced in the external portions than in the internal ones. This phenomenon is illustrated in Fig. 9.1, which is based upon the results of a bridge test on a long-span plate girder bridge [5]. This figure shows the

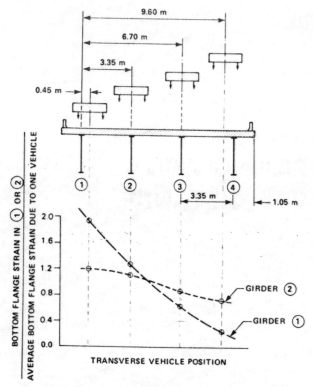

Figure 9.1 Observed bottom flange strains in girders.

variation of bottom flange strains in an external and an internal girder at midspan, due to the transverse positions of a vehicle. By interpolation of the plotted results it can be established that the strain in the external girder would be reduced by some 16 percent if the vehicle-edge distance were to be increased from 1.05 to 2.05 m (3.44 to 6.73 ft). This same increase in vehicle-edge distance would have little effect on the strain in the internal girder.

Developmental Background

The methods presented in this chapter to account for edge stiffening and increased vehicle-edge distance b_s, are intended to be used in conjunction with those for longitudinal moments given in Chap. 4. Their development, which is reported in Ref. 2, is based on the premise, shown in Fig. 9.2, that in an orthotropic plate two symmetrical portions which are outside the loaded area can be replaced without significant loss of accuracy by two symmetrical notional edge beams. The validity of this premise, within the practical range of bridges, has also been demonstrated in Ref. 2.

The method was developed to utilize the characterizing parameter λ, which is discussed in Sec. 1.9 and given by Eq. (2.34), and to introduce a modification factor F intended to be applied to the D values obtained by the methods of Chap. 4. Strictly speaking, the factor F depends upon a fairly large number of design parameters. However, in order to present the factor F in the form of charts and to keep the number of charts down to a manageable level, it was necessary to keep the number of variables as few as possible, consistent with retaining an accuracy sufficient for design purposes.

Following preliminary analyses by the orthotropic plate theory [3], the following conclusions were reached:

1. The value of F, while being dependent upon the number of lanes in a bridge, is little affected by variations in lane width. Accordingly, lane width considerations were dropped in the presentation of F values.

2. Variations in the torsion parameter α have a small, but not negligible, effect on F. It was concluded that, for the charts of F, the variations of α could be grouped into three zones, namely: 0.00 to 0.75, 0.76 to 1.49, and 1.50 to 2.00.

3. Variation of vehicle position transversely within a given lane has virtually no effect on F.

4. Values of F are quite insensitive to the number, weight, and spacing of the axles of the design vehicle.

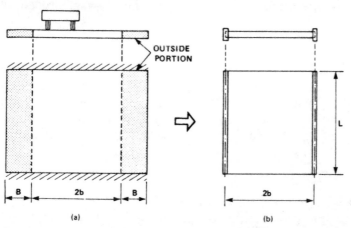

Figure 9.2 Notional edge beams. (*a*) Full orthotropic plate. (*b*) Outside portions replaced by notional edge beams.

5. The governing load cases (Figs. 4.10 and 4.18) remain unchanged by inclusion of the notional edge beams.

Longitudinal Shear

Unlike longitudinal moments, longitudinal shears are little affected by the transverse position of a vehicle. Therefore, very little advantage would be gained by modifying D values for shear, as obtained in Chap. 5, to account for an increased vehicle-edge distance. Also, because of the highly localized nature of longitudinal shear in the transverse direction, the beneficial influence of edge stiffening on the transverse distribution of longitudinal shear is usually not substantial. Accordingly, it is recommended that no account be taken of edge stiffening and increased vehicle-edge distance in the calculation of live-load longitudinal shears.

9.2 Method for Ontario Loading

The method for Ontario loading consists of reducing the actual edge-stiffening properties of the bridge, and the properties of that portion of the bridge which is represented by the vehicle-edge distance, to a pair of notional edge beams having flexural rigidity EI. Values of EI can be obtained according to the relevant procedure given in Sec. 2.5. From the known value of EI and other properties of the bridge, λ is calculated as follows:

$$\lambda = \frac{EI}{L}\left(\frac{1}{D_z^3 D_y}\right)^{0.25} = \frac{EI}{LD_z}\left(\frac{D_z}{D_y}\right)^{0.25} \tag{9.1}$$

where D_z and D_y are calculated according to the relevant procedure given in Sec. 2.2.

Method for ULS and SLS II

For longitudinal moments corresponding to the ultimate limit state and the serviceability limit state type II, the relevant method of Sec. 4.2 remains basically unchanged except that the following equation should be used instead of Eq. (4.7) for both external and internal portions.

$$D_d = \frac{D}{F}\left(1 + \frac{\mu C_f}{100}\right) \tag{9.2}$$

where the modification factor F is obtained from the relevant chart of Fig. 9.3 corresponding to the value of λ, calculated as above, and that of θ, calculated by the method of Sec. 4.2.

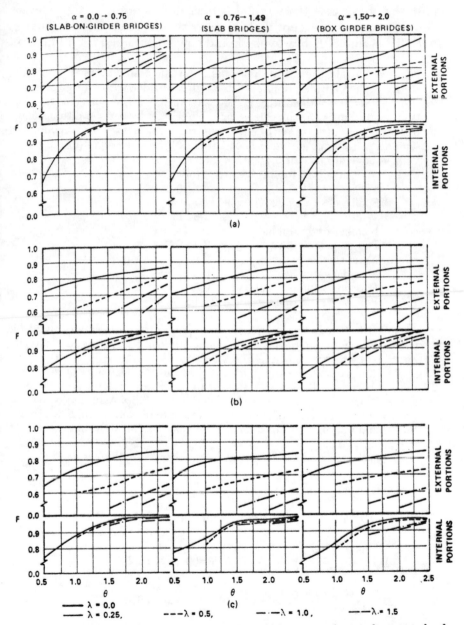

Figure 9.3 Modification factors corresponding to governing ultimate limit state load cases (OHBD code): (*a*) two-lane bridges; (*b*) three-lane bridges; (*c*) four-lane bridges. *Note: F = 1.0 for λ = 0.0.*

Method for SLS I

As in the above case the method is basically the same as the relevant method of Sec. 4.2, except that Eq. (9.2) should be used instead of Eq. (4.7). The value of F, however, should now be obtained from Fig. 9.4.

Example 9.1 The use of the above methods is illustrated by an example. The bridge under consideration is a slab-on-girder bridge with four girders, and the cross section of the bridge is shown in Fig. 9.5. It is required to calculate the longitudinal moment values for the one-lane-loaded condition for the following two cases: (a) With continuous barrier walls taken into account, and with a vehicle-edge distance b, of 1.0 m; (b) the same as in (a) but with a vehicle-edge distance of 2.70 m.

Relevant details of the bridge are as follows:

Span	= 24.7 m
Width	= 10.86 m
Number of design lanes	= 3
Design lane width	= 3.45 m
No of traveled lanes	= 2
Modular ratio	= 6.0
Poisson's ratio of steel	= 0.3
Poisson's ratio of concrete	= 0.15

Since the bridge is more than twice as long as it is wide, it is assumed that the girders and edge beams flex about a common straight neutral axis, the position of which is shown in Fig. 9.5a.

The longitudinal flexural rigidity of the bridge is calculated by disregarding the barrier wall, but with respect to the common neutral axis defined above. This calculation step is illustrated in Fig. 9.5b. Assuming a modular ratio of 6.0, D_x in terms of steel units is found to be $6.40 \times 10^6 E$, where E is the modulus of elasticity of steel and the quantity 6.4×10^6 has the units of mm^4/mm.

From the Poisson's ratios of steel and concrete, n_s is given by

$$n_s = \frac{6 \times 2(1 + 0.15)}{2(1 + 0.30)} = 5.3$$

D_y, D_{yx}, D_1, D_2, in steel and millimeter units, are obtained as follows, using Eq. (2.4):

$$D_y = \frac{E}{6.0}\left(\frac{190^3}{12}\right) \qquad\qquad = 0.95 \times 10^5 E$$

$$D_{yx} = \frac{G}{5.3}\left(\frac{190^3}{6}\right) = 2.16 \times 10^5 G = 0.83 \times 10^5 E$$

$$D_1 = 0.15 \times 0.95 \times 10^5 E \qquad = 0.14 \times 10^5 E$$

$$D_2 = D_1 \qquad\qquad\qquad\qquad = 0.14 \times 10^5 E$$

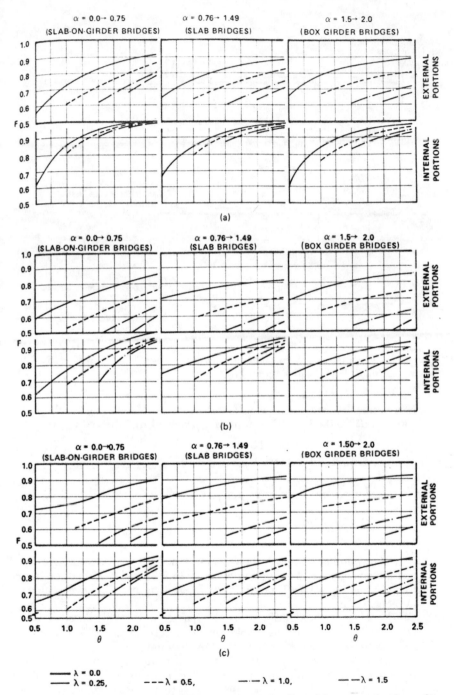

Figure 9.4 Modification factors for SLS I, Ontario loading: (*a*) two-lane bridges; (*b*) three-lane bridges; (*c*) four-lane bridges. *Note:* $F = 1.0$ for $\lambda = 0.0$.

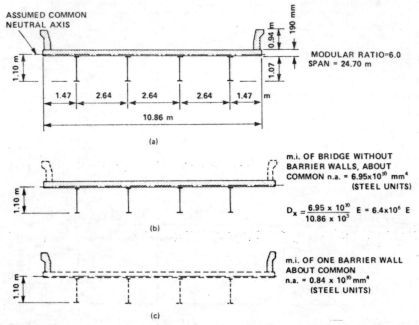

Figure 9.5 Calculation of flexural properties of a barrier wall: (*a*) cross section; (*b*) calculation of D_x; (*c*) calculation of I.

Neglecting the torsional inertia of steel girders, D_{xy} has the same value as D_{yx}, and is equal to $0.83 \times 10^5 E$. Hence from Eqs. (4.4), (4.5), and (4.6),

$$\alpha = \frac{2(0.83 + 0.14) \times 10^5 E}{2(64.0 \times 0.95)^{0.5} \times 10^5 E} = 0.12$$

$$\theta = \left(\frac{10.86}{2 \times 24.7}\right)\left(\frac{6.4 \times 10^6 E}{0.95 \times 10^5 E}\right)^{0.25} = 0.63$$

$$\mu = \frac{3.45 - 3.30}{0.6} = 0.25$$

Corresponding to the above values of α and θ, the D values for external and internal girders are found to be equal to 2.25 and 2.95 meters, respectively, from Fig. 4.13*b*. From the same figure C_f is found to have a value of 9 percent.

The second moment of area of the barrier wall is calculated about the common neutral axis of the whole superstructure, as shown in Fig. 9.5*c*; its value is found to be 0.84×10^{10} mm^4 in steel units.

The next step is to calculate the value of EI for the notional edge beam corresponding to condition (*a*) given above. For this condition the vehicle-edge distance is 1.0 m, which is the same as that considered in the development of the simplified methods of Chap. 4. Hence, in Eq. (2.35) EI_e is taken to be zero, that is, EI is taken to be $0.84 \times 10^{10} E$ (millimeter units).

Hence λ is obtained, using Eq. (9.1), as follows:

$$\lambda = \left(\frac{0.84 \times 10^{10}E}{24.7 \times 1000}\right)\left[\frac{1}{(6.4 \times 10^6E)^3(0.95 \times 10^5E)}\right]^{0.25} = 0.15$$

From Fig. 9.4, the charts for three-lane bridges give values of F of 0.78 and 0.80 for external and internal girders, respectively, for the above-calculated values of α, θ, and λ. Therefore, for external girders,

$$D_d = \frac{2.25}{0.78}\left(1 + \frac{0.25 \times 9}{100}\right) = 2.95 \text{ m}$$

Similarly, for internal girders,

$$D_d = \frac{2.95}{0.80}\left(1 + \frac{0.25 \times 9}{100}\right) = 3.77 \text{ m}$$

Turning to the second load case, i.e., condition (b), we investigate the effect of moving the vehicle to the middle of the traveled lane, which corresponds to a vehicle-edge distance of 2.70 m. This vehicle position is then as shown in Fig. 9.6a.

The curb width b_c is 0.25 m. Hence the flexural rigidity of the notional

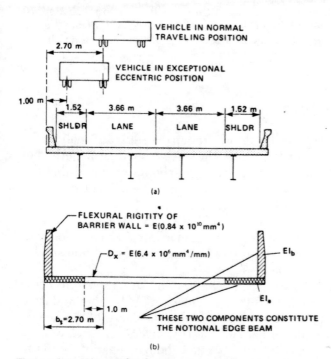

Figure 9.6 Notional edge beams incorporating barrier walls and vehicle-edge distance. (a) Bridge cross section. (b) Transverse distribution of longitudinal rigidity.

edge beam, EI_e, corresponding to the increased vehicle-edge distance is given by Eq. (2.38).

$$EI_e = 1000(2.70 - 1.0 - 0.25)(6.4 \times 10^6 E) = 0.93 \times 10^{10} E$$

Thence, from Eq. (2.35),

$$EI = (0.84 \times 10^{10} E) + (0.93 \times 10^{10} E) = 1.77 \times 10^{10} E$$

The two contributions to EI are shown diagrammatically in Fig. 9.6b.
From Eq. (9.1),

$$\lambda = \left(\frac{1.77 \times 10^{10} E}{24.7 \times 1000}\right)\left[\frac{1}{(6.4 \times 10^6 E)^3 (0.95 \times 10^5 E)}\right]^{0.25} = 0.32$$

For this value of λ, modification factors F for external and internal girders are found to be 0.58 and 0.60, respectively. Thus, for external girders,

$$D_d = \frac{2.25}{0.58}\left(1 + \frac{0.25 \times 9}{100}\right) = 3.97 \text{ m}$$

Similarly, for internal girders,

$$D_d = \frac{2.95}{0.60}\left(1 + \frac{0.25 \times 9}{100}\right) = 5.02 \text{ m}$$

It is interesting to note that increasing the edge distance to a realistic 2.70 m resulted in an increase in the D value for the internal girder from 2.95 to 3.97 m, or 34 percent. This means a reduction of live-load effects by 25 percent.

9.3 Method for AASHTO Loading

The AASHTO specifications [1] do not require a separate analysis for one-lane-loaded cases, and, therefore, there is no need for a method of analysis to deal with such cases. However, it should be noted that the method for SLS I analysis under Ontario loading (Sec. 9.2) can also be applied to a one-vehicle AASHTO loading.

For governing load cases, the method for AASHTO loading requires the same steps as those for ULS due to Ontario loading, given in Sec. 9.2, with the difference that Fig. 9.7 instead of Fig. 9.3 should be used for obtaining the values of the factor F.

This method could also be used in conjunction with the AASHTO D method [1]. However, in that case the F values should be obtained from the charts given in Fig. 9.8. It will be noted that the charts of Fig. 9.8 do not distinguish between external and internal girders. The charts are deliberately drawn this way in order to make them compatible with the AASHTO method, which itself does not distinguish between external and internal portions of a cross section. Drawing the charts of Fig. 9.8 in a

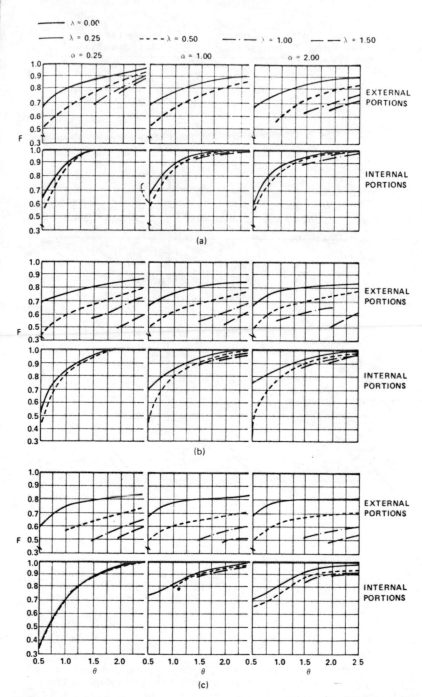

Figure 9.7 Modification factors corresponding to governing AASHTO cases: (a) two-lane bridges; (b) three-lane bridges; (c) four-lane bridges. *Note: F = 1.0* for $\lambda = 0.0$.

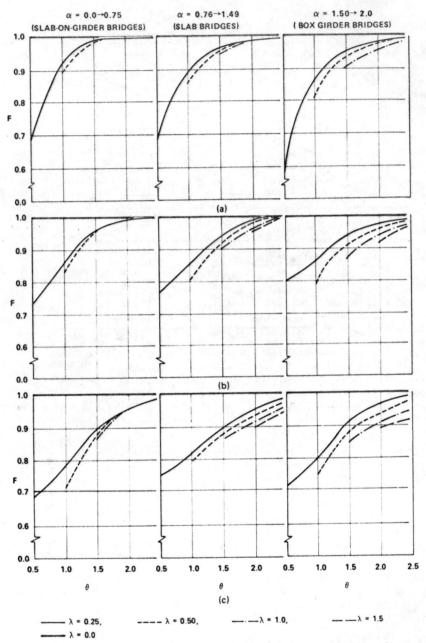

Figure 9.8 Modification factors corresponding to governing load cases for use with AASHTO D method: (*a*) two-lane bridges; (*b*) three-lane bridges; (*c*) four-lane bridges. *Note:* $F = 1.0$ for $\lambda = 0.0$.

way that does not distinguish between the two means that in any given case the value of F that is plotted is the larger, i.e., the more conservative, of the two values for external and internal portions. It follows that the limitations of the AASHTO method make it impossible to take full advantage of an increased vehicle-edge distance.

9.4   Moments in Edge Beams

An edge beam causes a reduction in live-load longitudinal moments in other portions of the bridge superstructure by attracting some of the moments to itself. An estimate of the live-load longitudinal moments taken by the edge beam can be obtained by the following.

As shown in Sec. 4.1, the live-load longitudinal moment M_g taken by a girder is given by

$$M_g = \frac{S}{D} M \tag{9.3}$$

where M is the moment due to one line of wheels. When edge beams are introduced, the above expression is revised as follows:

$$M_g = \frac{FS}{D} M \tag{9.4}$$

As can be seen in Fig. 4.3, M_g is also approximately equal to the product of the maximum intensity of longitudinal moment and girder spacing, that is,

$$M_g = M_{x,max} S \tag{9.5}$$

Equating (9.4) and (9.5),

$$\frac{FM}{D} = M_{x,max} \tag{9.6}$$

If the effect of transverse curvature on longitudinal moments is neglected, then $M_{x,max}$ is given by

$$M_{x,max} = -D_x \frac{\partial^2 w}{\partial x^2} \tag{9.7}$$

or $$\frac{\partial^2 w}{\partial x^2} = -\frac{M_{x,max}}{D_x} = \frac{-FM}{DD_x} \tag{9.8}$$

If the girder in question is an external girder, then it can be safely assumed that the longitudinal curvature of the edge beam is at least the same as that of the external girder, in which case the moment taken by the edge beam, M_e, is given by

$$M_e = -EI \frac{\partial^2 w}{\partial x^2} \tag{9.9}$$

whence, substituting the value of $\partial^2 w / \partial x^2$ from (9.8):

$$M_e = \frac{EI}{D_x} \frac{FM}{D} \tag{9.10}$$

Thus, if the values of EI, D_x, F, M, and D (for external portions) are known, the live-load moment in an edge beam can be approximately determined by Eq. (9.10). It is noted that the same equation can be derived for bridges other than slab-on-girder bridges. In fact, this same equation can be used for all shallow superstructures.

When an edge beam cannot be conceived as a separate beam, but is considered as part of the external girder or external portion, the live-load longitudinal moment taken by the combined edge beam and the external girder is given by $(FM/D)(S + EI/D_x)$.

References

1. American Association of State Highway and Transportation Officials (AASHTO): *Standard Specifications for Highway Bridges*, Washington, D.C., 1977.
2. Bakht, B. and Jaeger, L. G.: Effect of vehicle eccentricity on longitudinal moments in bridges. *Canadian Journal of Civil Engineering*, 10(4), 1983, pp. 582–599.
3. Cusens, A. R. and Pama, R. P.: *Bridge Deck Analysis*, Wiley, London, 1975.
4. Ministry of Transportation and Communications: *Ontario Highway Bridge Design Code (OHBDC)*, 2d ed., Downsview, Ontario, Canada, 1983.
5. Radkowski, A. F., Bakht, B., and Billing, J. R.: Design and testing of a 125 m span plate girder bridge. *International Conference on Short and Medium Span Bridges*, Toronto, Ontario, Canada, August 1982.

10

ANALYSIS OF CELLULAR AND VOIDED SLAB BRIDGES

10.1 Introduction

The small-deflection plate theory, which is widely employed for the analysis of bridges, is based upon the assumption that a plane section of the structure will remain plane after the application of the load. This assumption, while accurate enough for solid slab bridges, becomes open to question with cellular and voided slab bridges. The cross section of a cellular structure distorts because of the flexure of the flanges and webs about their own neutral axes. Because of this distortion, plane sections fail to remain plane after load application.

Cell Distortion

In the absence of frequently spaced diaphragms, a transverse slice of a cellular structure can be likened to a Vierendeel girder. It can be readily seen from Fig. 10.1 that a Vierendeel girder, because of the distortion of its cells, is much more flexible than its counterpart, in which the distortion is prevented by such means as cross bracings which provide the action of a truss. Frequently spaced transverse diaphragms in a cellular structure have the same effect as cross bracing in a Vierendeel girder, namely that of preventing the cell distortion.

In cellular structures lacking such frequently spaced diaphragms, the Vierendeel girder kind of behavior increases the effective flexibility of the transverse medium of the bridge which is responsible for transverse load distribution. This increase in flexibility cannot be accounted for by

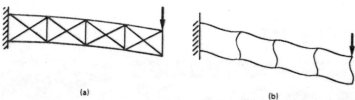

(a)   (b)

Figure 10.1 Effect of cell distortion on deflections: (a) a truss; (b) a Vierendeel girder.

reducing the flexural rigidity of the transverse medium, because the additional movement responds to applied shear rather than to the bending moment on the section. Hence, the effect of cell distortion is better represented by introducing a fictitious transverse shear area into the orthotropic plate idealization of the bridge.

It is noted that a solid slab does have an actual (i.e., physical) shear area. However, this actual shear area is large enough to have a negligible effect on load distribution in slabs, and hence in the usual orthotropic plate analysis shear deflections are ignored as compared with deflections due to bending. In effect, this assumes the shear area to be infinitely large and excluded from consideration.

The methods given in this chapter deal with the nature of adjustments that should be made to longitudinal moments and shears which have been obtained by methods that ignore the effect of cell distortion in the transverse direction, i.e., the methods of Chaps. 4 and 5. The methods developed in this chapter were recently developed in Ref. 2 through grillage analysis of idealized orthotropic plates which have finite shear areas in the transverse direction. The concept of analyzing such plates by grillage analysis is due to Sawko [4].

The analysis given in Ref. 2 shows that the local increase in intensity of longitudinal moments and longitudinal shear depends not only on θ and δ (as defined in Sec. 1.3), but also on the number of loaded lanes. Clearly, if all lanes of a bridge are loaded, all parts of the cross section tend to deflect almost equally, thus leaving little or no room for transverse cell distortion. In such cases, a shear-weak plate behaves like the one in which shear area is assumed to be infinite. Conversely, the effect of cell distortion on load distribution is most apparent when only one lane is loaded.

The number of loaded lanes which govern the design of a bridge at the ultimate limit state depends upon the modification factors for multilane bridges (Fig. 4.8). For bridges in which cell distortion is not a problem, the governing load cases are shown in Figs. 4.10 and 4.18. It has been confirmed by analysis that in these cases the introduction of "very shear weak" behavior, by postulating even the smallest realistic value of trans- verse shear area, does not alter which is the governing load case for a given bridge. Thus, the methods given in this chapter, except those for SLS I, are based upon the governing load cases shown in Figs. 4.10 and 4.18.

Limitations

The methods presented in this chapter, except those of Sec. 10.5, are to be used in conjunction with those of Chaps. 4 and 5 and are thus subject to the same limitations, which are given in Sec. 4.1.

For a cellular structure, the responses obtained from the methods of Chaps. 4 and 5 may be used without modification provided that there are transverse diaphragms, frequently spaced, or if the following conditions are met:

1. The web thickness is not less than 20 percent of the total depth of the section.

2. The void depth, t_v (Fig. 2.23) is not more than 80 percent of the total depth of the section.

If a structure satisfies the above conditions, its cell distortion effects are negligible, and therefore need not be accounted for.

Voided slab bridges which have circular or near-circular voids are less prone to cell distortion than cellular structures. It is shown in Ref. 2 that for most voided slab bridges the transverse cell distortion effects are so small that they can be safely neglected. For these bridges, analysis of cell distortion needs to be carried out only under the following conditions:

1. The void diameter is more than 80 percent of the total slab thick- ness, and/or

2. The center-to-center spacing of voids is less than the total slab thickness.

10.2 Longitudinal Moments, Ontario Loading

Details of the Ontario loading, lane widths, and dynamic load allowance are given in Sec. 4.2.

The following methods require the calculation of the parameter δ, which is given by Eq. (10.1) and calculated according to the relevant procedure of Sec. 2.4.

$$\delta = \frac{\pi^2 b}{L^2} \left(\frac{D_x}{S_y} \right)^{0.5}$$

(10.1)

The notation is defined in Chap. 2.

ULS and SLS II

The various steps in calculating longitudinal moments in cellular and voided slabs for ULS and SLS II are as follows:

1. Calculate D_d from Eq. (4.7) according to the method for ULS and SLS II for shallow superstructures as given in Sec. 4.2. It is recalled that, for cellular structures, approximate values of α and θ can be obtained from Eq. (2.12). Plate rigidities of voided slab bridges can be obtained from Eq. (2.9).

2. Calculate δ from Eq. (10.1). The value of transverse shear rigidity S_y for cellular structures can be obtained from Eq. (2.29) or (2.31) and that for a voided slab from Eq. (2.32).

3. Obtain the value of λ_m from the relevant chart of Fig. 10.2 corresponding to the value of δ, the value of θ, and the number of design lanes in the bridge.

4. Divide the value of D_d obtained in step 1 above by λ_m, to obtain the modified D_d value, which accounts for cell distortion effects and which should be used in design.

The value of λ_m is always larger than 1.0. Hence division of D_d by the modification factor always results in a smaller value of D_d, which in turn results in larger live-load effects. The physical interpretation of a value of λ_m close to 1.0 is that the transverse cell distortion effects are negligible.

SLS I

The SLS I loading differs from that for ULS or SLS II in that the former requires the loading of only one design lane, while the latter requires such loading as will produce the maximum load effects. As noted earlier, the cell distortion effects are maximized when only one lane of the bridge is loaded. Clearly, for such load cases the values of magnification factor λ_m will be larger than the corresponding values for the ultimate limit state, which usually requires the loading of more than one lane.

The method of accounting for the effects of cell distortion for SLS I is the same as for ULS and SLS II with the difference that the values of D_d

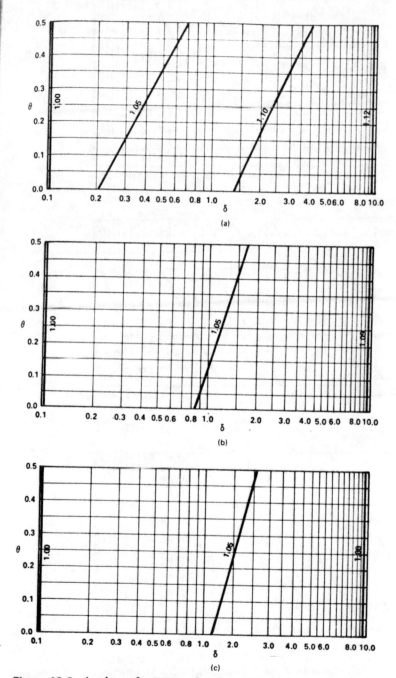

Figure 10.2 λ_m charts for ULS and SLS II, Ontario loading: (*a*) two-lane bridges; (*b*) three-lane bridges; (*c*) four-lane bridges.

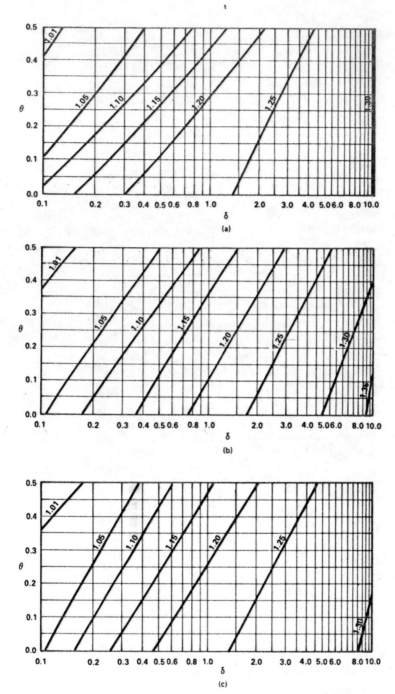

Figure 10.3 λ_m charts for one-lane-loaded cases (SLS I, Ontario loading): (a) two-lane bridges; (b) three-lane bridges; (c) four-lane bridges.

should be obtained for SLS I and the value of λ_m should be obtained from Fig. 10.3 instead of Fig. 10.2.

10.3 Longitudinal Moments, AASHTO Loading

Analysis for the case of one lane loaded is not required by AASHTO specifications. Thus, analysis is required only for the governing load case. This analysis is the same as for ULS due to Ontario loading, given in Sec. 10.2, with the difference that Fig. 10.4 should be used instead of Fig. 10.2 for obtaining the value of λ_m and the method for determining D_d should be that for AASHTO loading as given in Sec. 4.3.

Example 10.1 A single-span two-lane cellular bridge having a span of 90 ft (27.43 m) and the cross section as shown in Fig. 10.5 is required to be analyzed for live-load moments according to AASHTO loading.

Using Eq. (2.30), S_y is found to be equal to $0.08\,E_c$, and using Eq. (2.11), D_x is found to be equal to $10,816\,E_c$, both in inches. Thus from Eq. (10.1),

$$\delta = \pi^2 \frac{29(12)}{2(90 \times 12)^2} \left(\frac{10,816\,E_c}{0.08\,E_c} \right)^{0.5} = 0.54$$

Using Eq. (2.12), θ is found to be 0.16, and α is 1.0. For a lane width of 13.5 ft, μ is found to be 1.0 from Eq. (4.9).

From the method given in Sec. 4.3, D_d for both external and internal portions is found to be approximately 7.30 ft (2.22 m) for the above values of α, θ, and μ. For $\delta = 0.54$ and $\theta = 0.16$ the chart for two-lane bridges in Fig. 10.4 gives λ as 1.07. Therefore, the modified value of D_d for moments is $7.30/1.07 = 6.82$ ft (2.08 m).

10.4 Longitudinal Shears, Ontario and AASHTO Loadings

As discussed in Chap. 5, in those shallow superstructures in which transverse cell distortion effects are negligible, maximum longitudinal shears due to vehicle loads are, for all practical purposes, independent of θ values. This is not the case in cellular bridges, where cell distortion effects are substantial. For such structures, the transverse distribution of maximum longitudinal shears is characterized by α (which is approximately equal to 1.0), θ, and δ. The lack of dependence of longitudinal shear D values on θ for the structures without cell distortion can be seen in Table 10.1. This table lists the shear D values for slab bridges with different numbers of lanes, as obtained by grillage analysis for various load cases. For each case the D values are obtained for two span lengths, namely, 105 and 210 ft (32 and 64 m). Doubling the span results in the

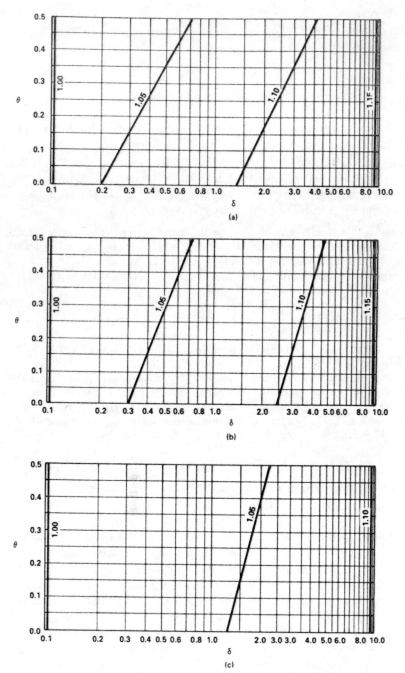

Figure 10.4 λ_m charts for AASHTO loading: (a) two-lane bridges; (b) three-lane bridges; (c) four-lane bridges.

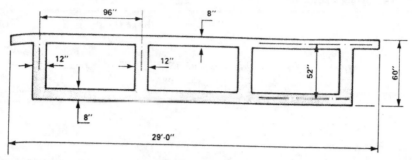

Figure 10.5 Cross section of a cellular bridge.

halving of the θ value. As can be seen in Table 10.1, even this significant change in θ has relatively little effect on the D values.

Table 10.2 shows the shear D values for the same cases, listed in Table 10.1, but in this case the structures are permitted to have transverse shear deformations. The shear areas of the transverse medium are so adjusted that δ is equal to 6.0 for both the 105- and 210-ft (32- and 64-m) span bridges. It can be seen in the table that the shear D values are no longer even approximately independent of the span. Therefore, it was found necessary to include θ in the analysis.

TABLE 10.1 Shear D Values for Slab Bridges with Negligible Transverse Shear Deformations as Obtained by Grillage Analysis

No. of lanes in the bridge	Bridge width, ft (m)	No. of loaded lanes	Shear D values, m	
			For bridge with 105-ft (32-m) span	For bridge with 210-ft (64-m) span
2	30.18 (9.20)	1	2.06	2.09
		2	1.73	1.76
3	39.37 (12.0)	1	2.28	2.27
		2	1.76	1.75
		3	1.68	1.69
4	52.49 (16.0)	1	2.55	2.61
		2	1.99	1.99
		3	1.76	1.76
		4	1.64	1.65

TABLE 10.2   Shear *D* Values for Cellular Bridges Having δ Equal to 6.0 as Obtained by Grillage Analysis

No. of lanes in the bridge	Bridge width, ft (m)	No. of loaded lanes	Shear *D* values, m	
			For bridge with 105-ft (32-m) span	For bridge with 210-ft (64-m) span
2	30.18 (9.20)	1	1.98	2.00
		2	2.00	2.00
3	39.37 (12.0)	1	1.98	2.07
		2	1.60	1.65
		3	1.57	1.68
4	52.49 (16.0)	1	1.92	2.12
		2	1.66	1.78
		3	1.50	1.59
		4	1.45	1.63

The Method

The simplified method for determining longitudinal shear due to live load in cellular bridges and voided slab bridges is basically the same as that for longitudinal moments as given in Secs. 10.2 and 10.3, with the difference that shear *D* values for the "no-transverse-distortion" case are obtained from either Sec. 5.2 or 5.3, whichever is relevant. These *D* values are then divided by the appropriate value of the modifier λ_s, which for the values of θ and δ of the bridge [Eqs. (2.12) and (10.1), respectively] is obtained from the relevant chart of Figs. 10.6 and 10.7.

It is noted that, as shown in Table 5.4, the number of loaded lanes governing longitudinal shears is the same for both AASHTO and Ontario loadings. Consequently, the values of the modifier λ_s are also the same for AASHTO loading as for the ULS and SLS II loading of the Ontario code.

Example 10.2 Example 10.1 for the two-lane bridge found that $\delta = 0.54$ and $\theta = 0.16$. From Table 5.3 the *D* value for shear for a slab bridge with two lanes is found to be 6.25 ft. From Fig. 10.6, the chart for two-lane bridges gives λ_s as approximately 1.03. Therefore, the *D* value for shear for the solid slab bridge can be modified to take account of transverse cell distortion of the cellular structure by dividing by 1.03. The final *D* value for shear is equal to 6.07 ft (1.85 m).

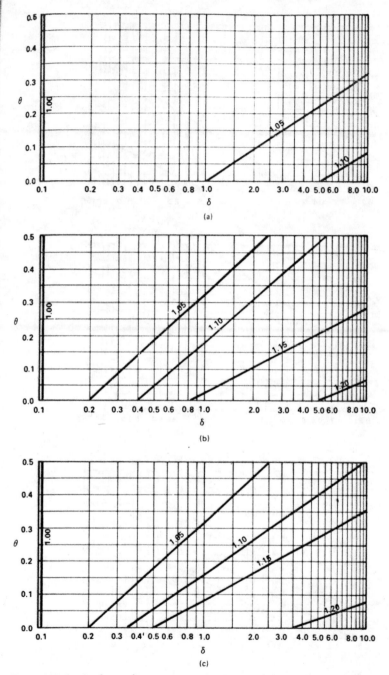

Figure 10.6 λ_s charts for AASHTO loading and for ULS and SLS II, Ontario loading: (*a*) two-lane bridges; (*b*) three-lane bridges; (*c*) four-lane bridges.

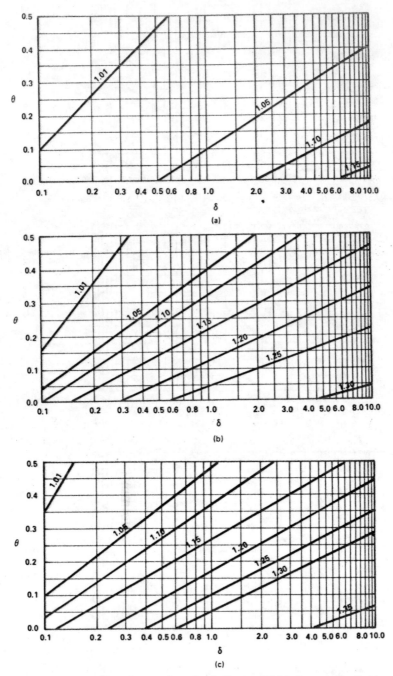

Figure 10.7 λ_s charts for one-lane-loaded cases (SLS I, Ontario loading): (a) two-lane bridges; (b) three-lane bridges; (c) four-lane bridges.

10.5   Voided Slab Bridge with a Central Isolated Support

The anatomy of the type of bridge dealt with in this section is shown in Fig. 10.8. The bridge is right, and has two equal spans with a central isolated support. As shown in the figure, it has a solid cross section for a

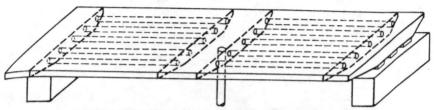

Figure 10.8   A voided slab bridge with an intermediate isolated support.

certain length near each abutment and over the intermediate support. For the remaining length, the slab incorporates longitudinal circular voids. There is a sufficient number of isolated supports at each abutment that, for the purpose of analysis, these supports can be collectively regarded as a single line support.

A complete simplified method of analysis has not yet been developed for the structure described above, even when the angle of skew is zero and the transverse cross section is fairly uniform across the width. The method given here deals with only the dead-load analysis of such a structure.

In the analysis for dead load of this kind of structure, it has been the usual practice to treat the structure as a two-span bridge. In this process, the transverse cross section containing the intermediate isolated column support is considered as unyielding. Such an approach gives an accurate assessment of the total longitudinal moments and shears at any cross section. However, the manner in which these responses are distributed across the width can be grossly in error.

As shown in Fig. 10-9a, the usual approach gives uniform intensities of longitudinal moments and shears. In practice these responses can be markedly nonuniform across a transverse cross section. The intensity of negative longitudinal moments in portions above the intermediate support is always significantly larger than the average intensity across the section. Since the total amount of longitudinal moment at a given cross section of a right bridge is invariant, this larger-than-average intensity must be accompanied by a smaller-than-average intensity at other points on the cross section. As shown in Fig. 10.9b, this smaller-than-average intensity of longitudinal moments occurs in the outer portions of the bridge.

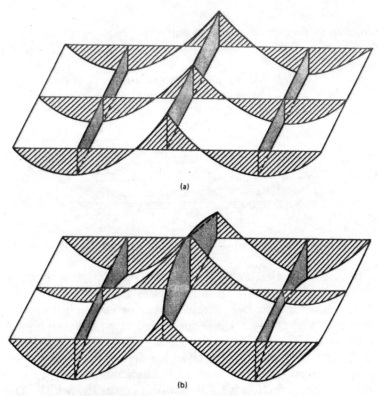

Figure 10.9 Contours of the intensity of longitudinal moments: (a) when intermediate support is continuous across the width; (b) when intermediate support is a single central column.

In the determination of the intensity of positive longitudinal moments, the pattern of error is reversed. In this case, as shown in Fig. 10.10, the larger-than-average intensity occurs near the outer portions, and the smaller-than-average intensity near the longitudinal centerline. It should be noted, however, that the distribution of longitudinal moments becomes less "peaky" away from the middle support; because of this phenomenon, the degree of error involved with the prediction of positive moments is negligibly small compared with that of the negative moments.

Longitudinal Moments

Longitudinal moments due to the uniformly distributed dead load in the bridge under consideration (Fig. 10.8) can be obtained by splitting them into two components:

1. Positive moments due to dead load in the slab supported at the two abutments, and without the intermediate column support.

2. Negative moments due to the intermediate column reaction, which for two equal spans is approximately $0.625W$, where W is the total dead weight of the two-span slab.

The former moments can be easily obtained by treating the bridge as a simply supported beam and assuming a uniform distribution of longitudinal moments across the bridge width. Thus, for example, if the distance between the abutments is $2L$ and the uniformly distributed dead load per unit area is w, then the moment intensity midway between the abutments is equal to $w(2L)^2/8$ per unit width, and this intensity displays a parabolic variation between the abutments.

The transverse profile of negative longitudinal moments due to the column reaction across the transverse section containing the column can be obtained with the help of charts in Fig. 10.11. This figure provides coefficients K_{m0}, K_{m1}, and K_{m2} for the negative moment intensities at midwidth, quarter width, and the longitudinal free edge of the slab, respectively. From these coefficients, the negative moments due to the column reaction are obtained as follows, with reference to the coordinate system shown in Fig. 10.11.

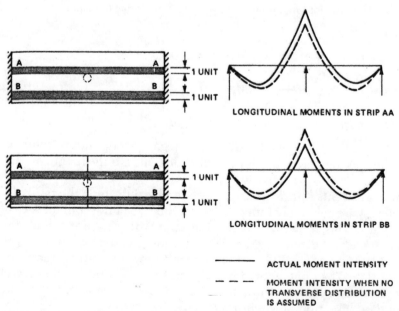

LONGITUDINAL MOMENTS IN STRIP AA

LONGITUDINAL MOMENTS IN STRIP BB

——— ACTUAL MOMENT INTENSITY

– – – MOMENT INTENSITY WHEN NO TRANSVERSE DISTRIBUTION IS ASSUMED

Figure 10.10 Errors in neglecting transverse distribution of dead-load longitudinal moments.

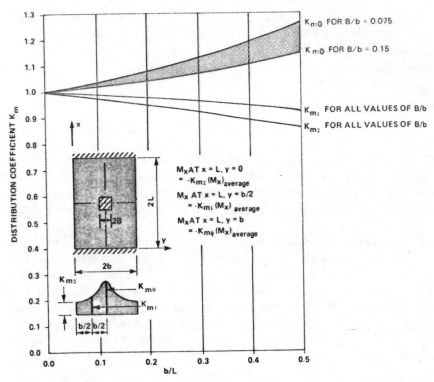

Figure 10.11 Chart for longitudinal moment coefficients due to column reaction.

At $x = L$, $y = 0$ or $2b$: $M_z = -K_{m2}M_{x,av}$

At $x = L$, $y = \dfrac{b}{2}$ or $\dfrac{3b}{2}$: $M_z = -K_{m1}M_{x,av}$ (10.2)

At $x = L$, $y = b$: $M_z = -K_{m0}M_{x,av}$

where $M_{x,av}$ is the average intensity of the negative longitudinal moment due to the column reaction, b is the half width, and L is the length of each span.

The coefficients K_{m0}, K_{m1}, and K_{m2} depend mainly upon the ratio b/L. K_{m1} and K_{m2} are relatively independent of the size of the contact area of the column. This, however, is not the case with K_{m0}, for which the column size effect cannot be ignored. Accordingly, the values of K_{m0} in Fig. 10.11 are given for different ratios of the width of contact area to the width of slab. The contact area is assumed to be square. In the case of rectangular contact areas, the actual dimension of the contact area in the longitudinal direction of the bridge should be used.

It is noted that the values of the moment coefficients were calculated from orthotropic plate theory analysis [3].

The values of negative longitudinal moments obtained by Eq. (10.2) and with respect to the coefficients given in Fig. 10.11 are applicable to the transverse section containing the intermediate column. Fortunately, rigorous analysis shows that bending moments at some distance from this section can be determined by the usual simplified technique described earlier without incurring any significant errors. Figure 10.12 shows the variation of K_{mo} in the longitudinal direction for two bridges with different aspect ratios. It can be seen that K_{mo} rapidly diminishes to values close to 1.0 as one moves away from the column, thus indicating that the longitudinal moments due to the central concentrated load are nearly uniformly distributed across transverse sections which are a distance of more than about $L/4$ from the column. It can be concluded that, as far as positive longitudinal moments are concerned, the practice of analyzing the bridge by assuming a rigid line support at the cross section containing the column support is justifiable for sections away from the column.

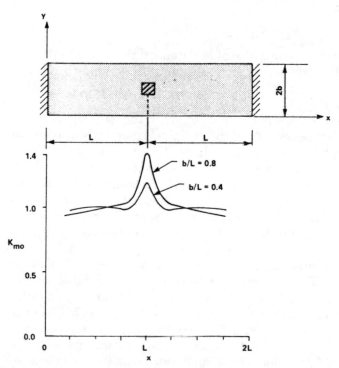

Figure 10.12   Variation of K_{mo} along the span.

Longitudinal Shears

In a similar manner to the treatment of longitudinal moments, the dead-load longitudinal shear intensities in the bridge under consideration (Fig. 10.8) can be obtained by splitting them into two components:

1. Shears due to the dead load in the bridge without the intermediate support.

2. Shears due to the column reaction alone.

The intensities of longitudinal shears of type 1 above are very nearly uniformly distributed across the cross section and are easily obtained by treating the bridge without the intermediate support as a simple beam and dividing the total longitudinal shear at a cross section by the bridge width.

Intensities of longitudinal shears of type 2 above can be obtained by the following equations, which make use of the distribution coefficients K_v given in Fig. 10.13.

$$\text{At } y = 0 \text{ or } 2b: \qquad V_x = K_{v2}V_{x,av}$$

$$\text{At } y = \frac{b}{2} \text{ or } \frac{3b}{2}: \qquad V_x = K_{v1}V_{x,av} \qquad (10.3)$$

$$\text{At } y = b: \qquad V_x = K_{v0}V_{x,av}$$

where $V_{x,av}$ is the average intensity of longitudinal shear due to the column reaction at the transverse section under consideration, and the reference axes are as shown in Fig. 10.13. Values of the distribution coefficient depend upon the ratio b/L.

Unlike the intensity of longitudinal moments, the intensity of longitudinal shear does not become even approximately uniformly distributed at transverse sections away from the column. Hence, it is necessary that the simplified method for shear be able to account for the nonuniform distribution of longitudinal shear throughout the length of the bridge. Accordingly, the values of K_v are given in Fig. 10.13 for transverse sections at $x = 0.25L$, $0.5L$, $0.75L$, and L', where L' is the distance between the support at the abutment and the nearer face of the column support.

For obtaining values of K_v at intermediate locations, graphical interpolation using three known values of K_v is recommended. For example, suppose that the value of K_{v0} is required at $x = 0.85L$ for a bridge with $b/L = 0.4$. From Fig. 10.13, values of K_{v0} for $x = 0.5L$, $0.75L$, and L' ($= 0.95L$, say) are 1.20, 1.42, and 2.16, respectively. These values are plotted in Fig. 10.14 with respect to x. From graphical interpolation the value of K_{v0} at $x = 0.85L$ is found to be approximately 1.64. It can be seen

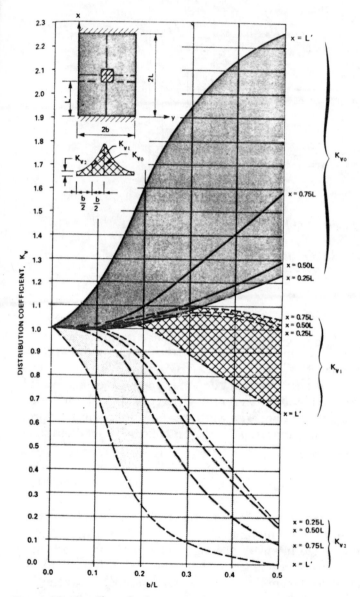

Figure 10.13 Chart for longitudinal shear coefficients due to column reaction.

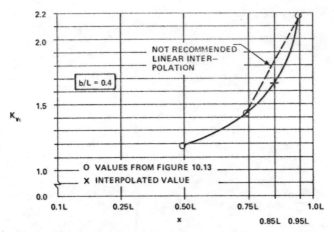

Figure 10.14   Interpolation of K_v.

in the figure that a linear interpolation would give a crude approximation; such, therefore, is not recommended in this case.

Transverse Moments

The dead-load transverse bending moments in the structure under consideration (Fig. 10.8) are mainly induced by the upward column reaction, and thus are negative, i.e., causing tension in the top fibers. The intensity M_y of this transverse moment depends upon the aspect ratio of the bridge represented by b/L and (in the vicinity of the column) upon the ratio of the column contact area width to the bridge width (represented by B/b). M_y at any point $(x, \bar{y})$ on the bridge can be obtained from the following equation:

$$M_y = -F \cos\left(\frac{\pi\bar{y}}{2b}\right)(M_{z,av}) \tag{10.4}$$

where the reference axes are as shown in Fig. 10.15, $\bar{y}$ is the distance of the reference point in the transverse direction from the bridge center line, $M_{z,av}$ is the average intensity of longitudinal moment due to the column reaction at the transverse section containing the reference point, and F is a coefficient read from Fig. 10.15 corresponding to the relevant values of x/L, b/L, and B/b.

Bridges with Nonuniform Cross Sections

Not many voided slab bridges are of constant depth across their entire transverse cross section. Most bridges have a uniform depth of construction in the middle portion and tapered outer portions, as shown in Fig. 10.8. The simplified methods of analysis described herein were basically developed for isotropic slabs in which the flexural and torsional rigidities are uniformly distributed across the width of the bridge. These methods can be readily modified so as to be applicable to bridges with tapered outer portions.

It has been shown in Ref. 1 that the outer portions of a rectangular orthotropic plate, simply supported along opposite edges, can be replaced by edge beams without any significant loss of accuracy, provided that the applied loads are away from the outer portions. In the case of an unstiffened orthotropic plate, the longitudinal rigidities are uniformly distributed in the outer portions, as they are in the remaining width. However, in the equivalent stiffened plate the outer-portion rigidities are lumped into edge beams. Hence it can be concluded that, for loads away from the outer portions, the manner of distribution of rigidities in the outer portions has little effect on transverse load distribution. It is noted that for the case under consideration the applied load, i.e., the upward column reaction, is well away from the tapered edges. Hence these tapered edges can be replaced by slab portions of reduced width having the same per-unit-width flexural rigidity as that in the middle portion. This transformation, as shown in Fig. 10.16, results in an equivalent bridge having constant depth and reduced width, which should be

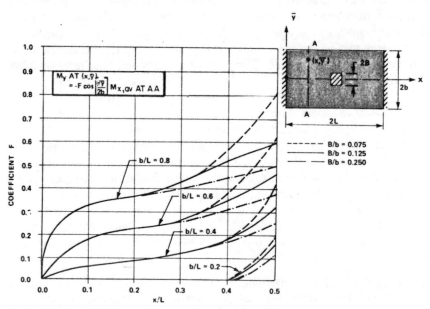

Figure 10.15 Coefficients for transverse moments due to column reaction.

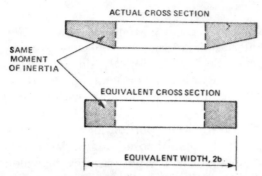

ACTUAL CROSS SECTION

SAME
MOMENT
OF INERTIA

EQUIVALENT CROSS SECTION

EQUIVALENT WIDTH, 2b

Figure 10.16 Idealizing variable cross section for dead-load analysis.

used in the calculations. It will be appreciated that in this transformation, the equivalence of longitudinal torsional rigidities is not maintained. However, the degree of error involved in such a lack of equivalence is small and can be neglected.

The calculation of positive moments corresponding to the structure without the intermediate column, and the interpretation of results from the simplified analysis of the bridge under the upward column reaction, require careful consideration.

The variation in moments of inertia per unit width of the tapered and uniform-depth portions of the cross section is much greater than the corresponding ratio of the dead load per unit width; i.e., the tapered and untapered portions of the cross section do not have the same ratio of moment of inertia to dead weight. The uniform-depth portions, being stiffer, carry more than their share of the dead weight, and the tapered portion less than their share. The load acceptance of the dead weights across the bridge width can be assumed to be in proportion to the moments of inertia per unit width of the various longitudinal strips, as demonstrated below in Example 10.3.

Analysis for the upward reaction can be readily carried out if the equivalent widths for the various longitudinal strips are separately established, as shown in the following example. It should be noted, however, that the approximations regarding the calculation of negative moments in tapered edges are valid only if the transverse curvatures are small. Since K_m varies between 0.90 and 1.20 for most practical cases (see Fig. 10.11), it can be concluded that the transverse curvatures are small and that the approximations employed in the following example are valid.

Example 10.3 To illustrate the use of the methods given here, a voided slab bridge with a central column is analyzed for dead-load effects. The plan of the

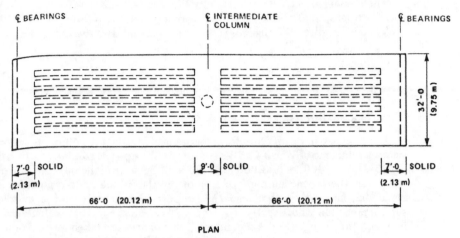

PLAN

Figure 10.17 Plan of a voided slab bridge.

bridge is shown in Fig. 10.17, and the cross section is shown in Fig. 10.18, which also shows equivalent slab widths for the various portions of the actual cross section. These various portions are identified in the figure as *A*, *B*, *C*, etc. It is assumed that the cross section has a common neutral axis. For each tapered portion, the equivalent width is obtained by dividing its second moment of area by the moment of inertia per unit width of portion *A*, which has a constant moment of inertia across its width. The equivalent widths are also shown in Fig. 10.18.

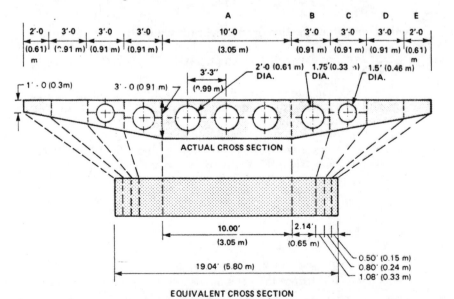

EQUIVALENT CROSS SECTION

Figure 10.18 Establishing equivalent cross section.

The contact area of the column with the slab is 2×2 ft (0.61×0.61 m), and the total weight of the voided portion of the bridge is 8400 lb/ft (122.6 kN/m) per length of the bridge. Hence the relevant data are as follows:

$L = 66.0$ ft (20.11 m)

$b = 9.52$ ft (2.90 m)

$B = 1.0$ ft (0.30 m)

$w = 8400$ lb/ft (122.6 kN/m)

For a two-span beam of uniform moment of inertia, the column reaction under a total load of W is $0.625W$. Therefore, the upward column reaction is $0.625(2)(66)(8400) = 693,000$ lb (3083 kN). Because of the introduction of the voids, the second moment of area of the slab in the middle portions is somewhat larger than that near the supports. In such a case the middle support reaction is not exactly equal to $0.625W$. A more accurate calculation, in which the changes in the second moments of area are taken into account, leads to a middle-support reaction of 695,000 lb (3092 kN), an increase of only 0.3 percent.

It is required to calculate longitudinal moments in the middle and outer strips, that is, A and E, respectively.

The total dead weight of the structure is 8400 lb/ft (122.6 kN/m), and total moment of inertia of the cross section is 42.3 ft^4 (0.364 m^4). The average moments of inertia of strips A and E are readily found to be 2.22 ft^4/ft (0.062 m^4/m) and 0.55 ft^4/ft (0.016 m^4/m), respectively. Therefore, the effective dead weight for strip A is

$$\frac{2.22}{42.3}(8400) = 441 \text{ lb/ft (6.44 kN/m)}$$

and similarly the effective dead weight for strip E is

$$\frac{0.55}{42.3}(8400) = 109 \text{ lb/ft (1.59 kN/m)}$$

The dead-load moments per unit width in strips A and E of the structure without the intermediate column are shown in Fig. 10.19, together with the average longitudinal moments due to the upward column reaction of 693,000 lb (3083 kN). It is noted that the moments due to this column reaction are averaged over the equivalent width of 19.04 ft (5.80 m).

The two geometric ratios required for the analysis are as follows:

$$\frac{b}{L} = \frac{19.04}{2(66)} = 0.14$$

$$\frac{B}{b} = \frac{1.0(2)}{19.04} = 0.11$$

From Fig. 10.11, K_{m0} and K_{m2} for these ratios are found to be 1.03 and 0.97, respectively. For strip A the net amount intensity over the column is

$$[960.5 - 1.03(1201.2)](1000) = -276,600 \text{ lb} \cdot \text{ft/ft } (-1230.4 \text{ kN} \cdot \text{m/m})$$

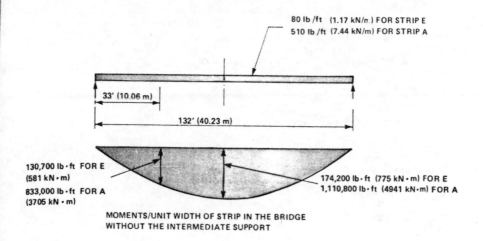

80 lb /ft (1.17 kN/m.) FOR STRIP E
510 lb /ft (7.44 kN/m) FOR STRIP A

33' (10.06 m)

132' (40.23 m)

130,700 lb·ft FOR E
(581 kN·m)

833,000 lb·ft FOR A
(3705 kN·m)

174,200 lb·ft (775 kN·m) FOR E
1,110,800 lb·ft (4941 kN·m) FOR A

MOMENTS/UNIT WIDTH OF STRIP IN THE BRIDGE
WITHOUT THE INTERMEDIATE SUPPORT

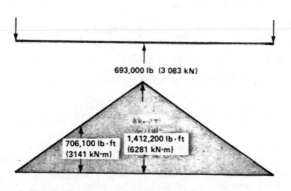

693,000 lb (3 083 kN)

706,100 lb·ft
(3141 kN·m)

1,412,200 lb·ft
(6281 kN·m)

AVERAGE MOMENT PER UNIT WIDTH DUE TO THE COLUMN
REACTION [AVERAGED OVER THE EQUIVALENT WIDTH OF
16.24 FT (4.95 m)]

Figure 10.19 Two components of average longitudinal moments.

Midway between the abutment and the column, the moment intensity is

$$(720.4 - 600.5)(1000) = 119,800 \text{ lb} \cdot \text{ft/ft} \ (532.9 \text{ kN} \cdot \text{m/m})$$

For strip E, an actual width of 2.0 ft (0.61 m) corresponds to an equivalent width of 0.50 ft (0.15 m). Therefore, the moments due to column reactions obtained from the simplified analysis should be multiplied by the fraction $0.50/2.0 = 0.25$ to obtain the actual intensity of moments. For strip E, the longitudinal moment at the cross section containing the column is

$$237,400 - 0.97(0.25)(1,201,100) = -53,866 \text{ lb} \cdot \text{ft/}$$
$$\text{ft} \ (-239.6 \text{ kN} \cdot \text{m/m})$$

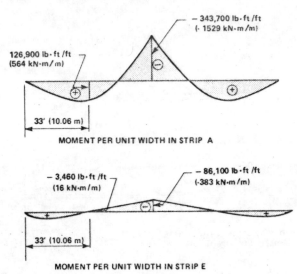

MOMENT PER UNIT WIDTH IN STRIP A

MOMENT PER UNIT WIDTH IN STRIP E

Figure 10.20   Net moments due to dead load.

The moment midway between the abutment and the column is

$$[178.1 - 0.25(600.6)](1000) = 27,950 \text{ lb} \cdot \text{ft/ft } (124.3 \text{ kN} \cdot \text{m/m})$$

The moment diagrams for the two strips are shown in Fig. 10.20.

References

1. Bakht, B., and Jaeger, L. G.: Effect of vehicle eccentricity on longitudinal moments in bridges, *Canadian Journal of Civil Engineering*, 10(4), 1983, pp. 582–599.
2. Bakht, B., Jaeger, L. G., and Cheung, M. S.: Cellular and voided slab bridges, *Journal of the Structural Division*, ASCE, 107(ST9), 1981, pp. 1797–1813.
3. Cusens, A. R., and Pama, R. P.: *Bridge Deck Analysis*, Wiley, London, 1975.
4. Sawko, F.: Discussion on paper by O. A. Kerensky, *The Structural Engineer*, 46(7), 1968, p. 204.

11

ANALYSIS OF MULTIBEAM AND MULTISPINE BRIDGES

11.1 Introduction

In a slab-on-girder bridge the transfer of load from a loaded girder to adjacent girders takes place mainly through the flexural rigidity of the deck slab. By contrast, a multibeam bridge, as shown in Fig. 2.18, may have only a very thin deck slab. In such a case the transverse load distribution takes place mainly through the shear keys which are between the beams.

If a multibeam bridge is idealized as an orthotropic plate, then it can be readily appreciated that both its transverse flexural rigidity D_y and the transverse torsional rigidity D_{yx} will be effectively zero. The analysis of this special case of the orthotropic plate, called the *articulated plate*, is the basis of the simplified methods presented in this chapter.

Articulated Plates

A scrutiny of Eqs. (2.1) and (2.2) will readily show that both the "normal" orthotropic plate parameters α and θ tend to infinity as D_y tends to zero, and hence that $1/\alpha$ and $1/\theta$ both approach zero. This means that if, temporarily, the space of values of $1/\alpha$ and $1/\theta$ is employed, one approaches the single point $(0, 0)$. This in turn suggests that a single parameter should be capable of characterizing the behavior. In Sec. 1.6 this parameter has been developed as $\theta/\sqrt{\alpha}$.

The load distribution methods of rectangular articulated plates which are simply supported along two opposite edges have been developed in Ref. 4 by means of a single parameter β defined as follows:

$$\beta = \pi \left(\frac{2b}{L}\right)\left(\frac{D_x}{D_{xy}}\right)^{0.5} \tag{11.1}$$

where the notation is as defined in Chap. 2. It will be apparent that, since $\theta/\sqrt{\alpha}$ is $\sqrt{2}\,(b/L)\,(D_x/D_{xy})^{0.5}$, the parameter β is the same as $\theta/\sqrt{\alpha}$ to within a simple constant multiplier. Hence, either β or $\theta/\sqrt{\alpha}$ may be used, at choice, as the characterizing parameter.

It is instructive to examine the physical significance of the parameter β with reference to Fig. 11.1, which shows the cross section of a multibeam bridge idealized as an articulated plate. In the case in which only one beam is loaded by a concentrated load it can readily be appreciated that, as the width $2b$ increases, the transverse load distribution becomes more and more uneven (that is, worse). As the width of the bridge tends to infinity, the value of β also tends to infinity. Thus, the larger the value of β the worse the load distribution. Conversely, when the span length increases, β becomes small and the load distribution characteristics of the bridge improve. Clearly, if L becomes very large, the bridge tends to behave as a beam, in which case concentrated loads, irrespective of their transverse positions, will be almost evenly distributed across the width. Hence, as β approaches zero, the bridge behavior approaches that of a beam.

The quantity D_x corresponds to the familiar flexural rigidity EI of a beam, and similarly D_{xy} corresponds to the familiar torsional rigidity GJ of the beam. Consider the case of Fig. 11.1 in which only one beam is

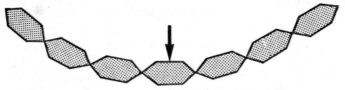

Figure 11.1 Cross section of an idealized articulated plate.

loaded; it can readily be seen that an increase in the flexural rigidity EI of the beams will result in the loaded beam taking a larger share of the load, so that the transverse load distribution pattern will worsen. Hence an increase of D_x should increase β, as indeed it does. If the EI of the beams becomes infinite, the deflections of all beams, including the loaded beam, will be zero, and consequently the loaded beam alone will sustain all the applied load, with nothing being transferred to other beams. In such a case β will be equal to zero.

Multispine Bridges

When the transverse flexural rigidity D_y of an orthotropic plate is very small compared with its longitudinal flexural rigidity D_x, the plate behaves substantially like an articulated plate, and the actual value of D_y becomes irrelevant as far as the load distribution is concerned. When D_y

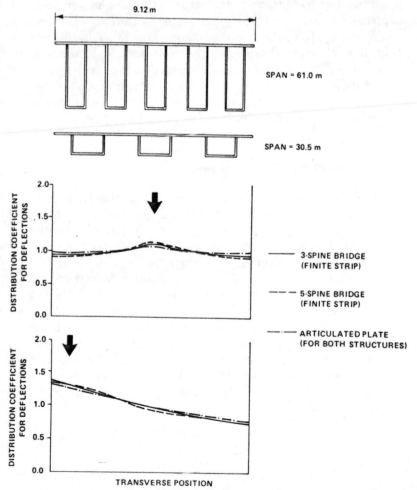

Figure 11.2   Distribution coefficients for deflections.

is smaller than, say, $0.05\,D_x$, then for all practical purposes the orthotropic plate can be treated as an articulated plate.

The transverse flexural and torsional rigidities of multispine bridges are usually small compared with their respective counterparts in the longitudinal direction. Therefore, these structures can realistically be idealized as articulated plates.

To verify that multispine bridges can be idealized as articulated plates, and hence characterized by β, the behavior of two different multispine bridges is examined. As shown in Fig. 11.2, one of the bridges has five spines and a span of 61 m (200 ft) and the other has only three spines and a span of 30.5 m (100 ft). The properties of the two structures are so adjusted that they both have β equal to 1.16. The two bridges were analyzed under single concentrated loads by the finite strip method, which makes it possible to idealize the structure as a three-dimensional assembly of plates. The idealized articulated plates for the two structures were also analyzed by the articulated plate method [3]. Distribution coefficients for deflections as obtained by the various analyses are compared with each other in Fig. 11.2. It can be seen that the results of the various analyses are imperceptibly different from each other. Other similar comparisons are given in Ref. 1, confirming the validity of the premise that multispine bridges can be analyzed as articulated plates.

11.2   Methods for Longitudinal Moments

It has been established that the load distribution in a bridge with finite but small values of D_y can be characterized by a single parameter β, or, equivalently, by the parameter $\theta/\sqrt{\alpha}$. This, however, does not negate the fact that the load distribution of the bridge can still be characterized by α and θ if desired. Figure 11.3 is very instructive in this regard. Plotted on Fig. 11.3 is the contour $\beta = 1.16$, which is the same as $\theta/\sqrt{\alpha} = 0.261$. It can be seen that this contour $\beta = $ constant is parallel to the adjacent D-value contours, and hence is itself a contour of D. This confirms that a change in D_y, while changing both α and θ, has no effect on D, that is, no effect on the load distribution pattern, in this region of the (α, θ) space. On the basis of the above, the methods for longitudinal moments for shallow superstructures as given in Chap. 4 can be used with minimal modifications, as shown below. The following equation is readily obtained from Eqs. (1.19) and (1.20).

$$\beta = \sqrt{2}\,\pi \cdot \frac{\theta}{\sqrt{\alpha}} \tag{11.2}$$

Knowing D_x and D_{xy} we may calculate the value of β from Eq. (11.1). We neither know, nor need to know, D_y and D_{yx}.

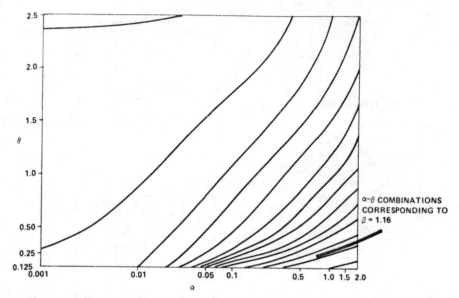

CHART FOR LONGITUDINAL MOMENTS IN 3-LANE BRIDGES WITH TWO LANES LOADED

Figure 11.3 α, θ curve corresponding to one β value.

Choosing a suitable value of α, say $\alpha = 2.0$, we then calculate θ from Eq. (11.2) as $\theta = \beta/\pi$. Then entering the (α, θ) chart at the point $\alpha = 2.0$, $\theta = \beta/\pi$, we find the value of D in the usual way. The choice of $\alpha = 2.0$ is arbitrary. Any other high value of α would serve equally well. It is noted, however, that the articulated plate behavior falls into that portion of the (α, θ) space in which α takes high values; in Fig. 11.3 the contours of D become well behaved, and in fact also become contours of β on the right-hand part of the chart, say, in the region $\alpha \geqslant 1.0$.

Once the value of D has been obtained, calculation of live-load longitudinal moments in multibeam bridges and multispine bridges proceeds in the manner given in Chap. 4.

Alternatively, the D values for longitudinal moments can be obtained directly from the charts given in Fig. 11.4. These charts give D values for longitudinal moments for two-, three-, and four-lane bridges for both AASHTO and OHBDC loadings. For AASHTO loadings the charts give D values for the governing load cases; for OHBDC loadings curves are given for one-lane-loading cases (SLS I), and for the ULS and SLS II cases. Correction factors C_f, similar to those used in the methods for shallow superstructures, are also given in Fig. 11.4. The design D value, that is, D_d, is obtained from the following equation.

$$D_d = D\left(1 + \frac{\mu\, C_f}{100}\right) \qquad (11.3)$$

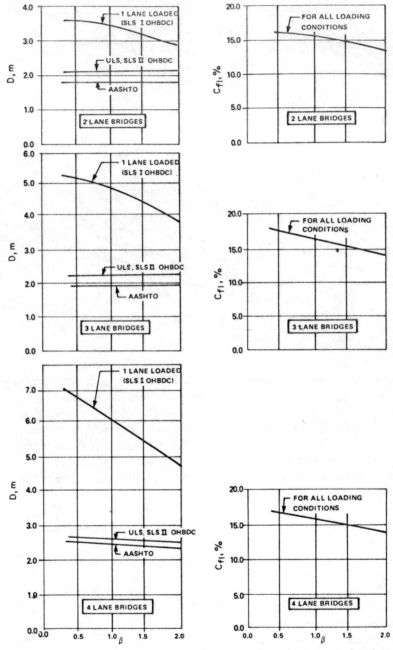

Figure 11.4 *D* values for longitudinal moments for multispine and multi-beam bridges.

where

$$\mu = \frac{W_e - 3.3}{0.6} \qquad \mu > 1.0 \tag{11.4}$$

with the design lane width W_e in meters. If W_e is in feet, then μ is obtained from the following expression:

$$\mu = \frac{W_e - 11.0}{2.0} \qquad \mu > 1.0 \tag{11.5}$$

It should be noted that the D_d values thus obtained are applicable to both internal and external portions of the cross section. If different values are required for external and internal portions, then the earlier-described adaptation to the method for shallow superstructures should be used.

11.3   Method for Longitudinal Shears

It is usual, but incorrect, to analyze bridges for both longitudinal moments and longitudinal shears using the same distribution factors. For example, according to AASHTO specifications, the same D value is used for longitudinal shears as for longitudinal moments. As discussed elsewhere, the transverse distribution of the higher derivatives of deflections becomes more and more localized. Thus, the transverse distribution of longitudinal shears is more localized, or "peakier," than that of longitudinal moments. (It is recalled that the former is related to the third derivatives of deflections, and the latter to the second.) Consequently, the D value for longitudinal shears may be expected to be smaller than the corresponding value for longitudinal moments.

The Ontario highway bridge design code [2] specifies that for multispine bridges the value of D for longitudinal shear should be taken as 0.9 times the corresponding value for longitudinal moments. The same criterion can be applied to obtain the shear D values for multibeam bridges.

It should be appreciated that the difference between the D values for shear and moment diminishes as the number of loads across the bridge width increases. Therefore, strictly speaking, the multiplier for obtaining shear D values from moment values should be smaller for one-lane-loaded cases than that for multilane loading. The multiplier given is realistic for one-lane-loaded cases but somewhat conservative for multilane loading.

Example 11.1   A single-span, two-spine bridge is analyzed here. The bridge has a span of 42.7 m (140 ft) and is composed of steel spines and a composite concrete deck slab. Details of the cross section are shown in Fig. 11.5. The

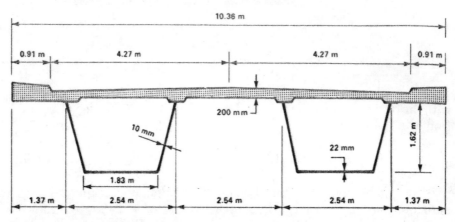

Figure 11.5 Cross section of a multispine bridge.

modular ratio for moduli of elasticity of steel and concrete is assumed to be 8.0. Thus, according to Sec. 2.3, the modular ratio n_s for shear moduli is equal to $0.88 \times 8 \, (= 7.04)$. It is required to calculate the D_d values for Ontario loading for both longitudinal moments and longitudinal shears.

From the relevant equations of (2.23) D_x is calculated to be equal to $0.224\,E_c$ in metric and concrete units. Also from the relevant equation of (2.23), D_{xy} is obtained as follows:

$$D_{xy} = \frac{G_c}{5.08} \frac{4(3.76)^2}{(1/7.04)\,[1.83/0.022 + 2(1.66)/0.010] + 2.54/0.20}$$

$$= 0.181\,G_c$$

β is given by Eq. (11.1):

$$\beta = \frac{10.36 \times \pi}{42.7}\left(\frac{0.224E_c}{0.181G_c}\right)^{0.5}$$

For Poisson's ratio of 0.15, $E_c = 2.3\,G_c$. Thus β is calculated to be equal to 1.29. For this value of β the charts for two-lane bridges, as given in Fig. 11.4, give D values for SLS I and for ULS as 3.30 m and 2.10 m, respectively. C_A is found to be 15 percent. From Eq. (11.4),

$$\mu = \frac{4.27 - 3.3}{0.6} = 1.62 \qquad \text{but not greater than 1.0}$$

Hence $\mu = 1.00$ is used. D_d for SLS I is given by

$$D_d = 3.30\left[1 + \frac{1.00(15.0)}{100}\right] = 3.80 \text{ m}$$

D_d for ULS and SLS II is given by

$$D_d = 2.10\left[1 + \frac{1.00(15.0)}{100}\right] = 2.42 \text{ m}$$

For a two-spine bridge, the distinction between internal and external portions is possible if each web of each spine is considered as a separate beam. The values of D_d calculated above are the more severe of the values for external and internal portions. If, for the ultimate limit state, it is desired to obtain separate values of D_d for internal and external portions, then the other approach given in Sec. 11.2 should be used. According to this approach, $\alpha = 2.0$ and θ is given from Eq. (11.2) as follows:

$$\theta = \frac{\sqrt{2.0}(1.29)}{\sqrt{2.0}\,\pi} = 0.41$$

For these values of α and θ the charts of Fig. 4.11b give D for external and internal portions as 2.10 and 2.05 m, respectively, and C_R as 18 percent. Thus, D_d for external portions is given by

$$D_d = 2.10\left[1 + \frac{1.0(18)}{100}\right] = 2.48 \text{ m}$$

and D_d for internal portions is given by

$$D_d = 2.05\left[1 + \frac{1.0(18)}{100}\right] = 2.41 \text{ m}$$

The fact that the smaller of these two D_d values is equal to the D_d value obtained by the other approach confirms that the two approaches give similar results.

The D_d value for longitudinal shear is obtained simply by multiplying the D_d values for moment by 0.9. Thus, for example, the shear D_d value for SLS I is equal to 0.9×3.80, or 3.42 m.

References

1. Cheung, M. S., Bakht, B., and Jaeger, L. G.: Analysis of box-girder bridges by grillage and orthotropic plate methods, *Canadian Journal of Civil Engineering*, 9(4), 1982, pp. 595–601.
2. Ministry of Transportation and Communications: *Ontario Highway Bridge Design Code (OHBDC)*, 2d ed., Downsview, Ontario, Canada, 1983.
3. Pama, R. P., and Cusens, A. R.: Edge stiffening of multi-beam bridges, *Journal of the Structural Division ASCE*, 93(ST2), 1967, pp. 141–161.
4. Spindel, J. E.: A study of bridge slabs having no transverse flexural stiffness, Ph.D. thesis, London University, London, 1961.

ANALYSIS OF FLOOR SYSTEMS OF TRUSS AND SIMILAR BRIDGES

12.1 Introduction

Chapters 1 to 11 have dealt basically with those superstructures in which the main longitudinal discrete members, if present, are four or more in number. This chapter deals with the analysis of a fundamentally different type of superstructure, that of a floor system supported on two main longitudinal members. The floor system may be transversely contained within the two longitudinal members, as, for example, in through-truss bridges. Alternatively, the floor system may project transversely beyond the longitudinal members, as is the case of a deck truss bridge. Various examples of the two types of floor systems are shown in Fig. 12.1.

The two types of floor systems identified above are usually employed for fairly long spans, as a result of which the effective flexural rigidities of the main longitudinal members are relatively large compared with those of other components in the system. For the purpose of analyzing a floor system in this kind of superstructure, it can safely be assumed that the main longitudinal members are nondeflecting. The finite flexural rigidity of these longitudinal members tends to improve load distribution in the floor system, and therefore the neglect of it leads to safe-side results. It

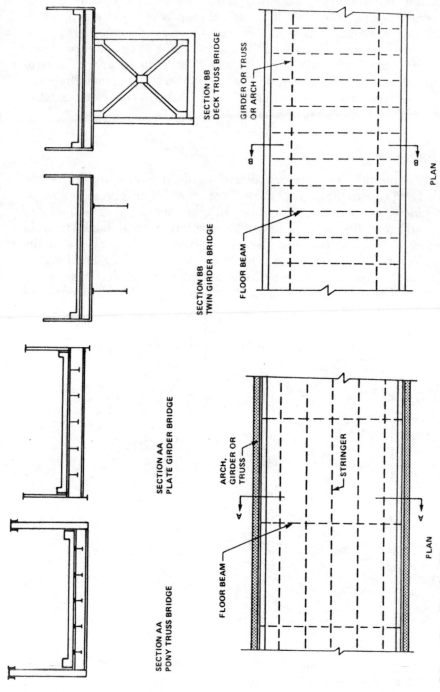

Figure 12.1 Details of floor systems of truss, arch, and twin-girder bridges.

SECTION AA
PONY TRUSS BRIDGE

SECTION AA
PLATE GIRDER BRIDGE

SECTION BB
TWIN GIRDER BRIDGE

SECTION BB
DECK TRUSS BRIDGE

GIRDER OR TRUSS
OR ARCH

FLOOR BEAM

PLAN

ARCH,
GIRDER OR
TRUSS

FLOOR BEAM

STRINGER

PLAN

should be noted that the amount of error which results from neglect of the deflections of the main longitudinal members is extremely small. Hence the approach, although conservative, is not excessively so.

The floor system of a truss or similar type of bridge spans predominantly in the transverse direction. This contrasts with the floor systems of other bridges which span predominantly in the longitudinal direction. This difference in the orientation of the span direction necessitates separate simplified analysis procedures.

Nomenclature

The longitudinal and transverse directions of the floor systems covered in this chapter are the same as those for other superstructures, in spite of the difference in span orientations. In other words, *longitudinal* still means the direction of traffic flow. This orientation is maintained to avoid the possibility of any confusion. Transverse beams spanning between the main longitudinal members are referred to as *floor beams*, and longitudinal beams spanning between the floor beams are referred to as *stringers*. The continuous riding surface may be provided by a concrete

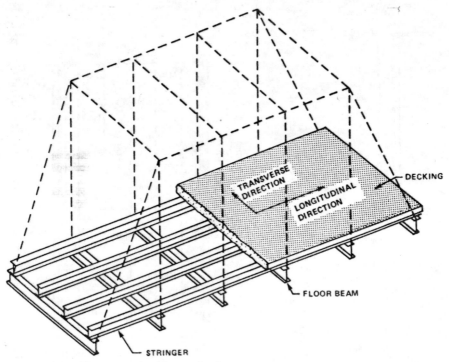

Figure 12.2 Floor system of a truss bridge.

slab, a steel grating, a laminated-wood deck, or timber planks; these components are generically referred to as *decking*. The notation for the various components of the floor system is illustrated in Fig. 12.2.

12.2   Analysis of Decking

As shown in Fig. 12.3*a*, a floor system may be comprised of only floor beams and no stringers, in which case the decking is supported directly by the floor beams and spans predominantly in the longitudinal direction. Alternatively, as shown in Fig. 12.3*b*, the floor system may include stringers, in which case the decking is supported by the stringers and spans predominantly in the transverse direction. The two cases require separate simplified analysis procedures for the decking.

Decking Supported Directly by Floor Beams

For the case where the decking is supported directly on floor beams, as shown in Fig. 12.3*a*, it is sufficiently accurate to assume that the floor beams are nondeflecting, and to analyze the decking as a slab bridge, a timber-laminated bridge, or a similar type of bridge, as the case may be, spanning between adjacent floor beams as rigid supports. Thus, longitudinal moments and shears due to live load in the decking can be obtained by the relevant D-type method of Chaps. 4 and 5. The span of the decking, i.e., the spacing of the floor beams, should, of course, be taken as the span of the conceptual local bridge. It is noted that, for continuous decking, modifications to the span length will be necessary for calculating longitudinal moments. These modifications, which are required only for the calculation of θ [Eq. (4.5)], should be carried out in accordance with the procedures given in Chap. 8.

Transverse moments due to live loads in concrete deck slabs can be calculated according to the method given in Sec. 6.3.

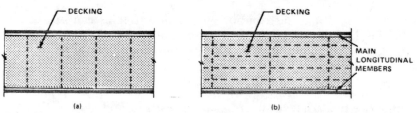

Figure 12.3 Partial plans of floor systems: (*a*) decking supported on floor beam only; (*b*) decking supported on stringers.

Decking Supported by Stringers

When the decking is supported on stringers, as shown in Fig. 12.3b, the system comprising decking plus stringers can be regarded as a conceptual local bridge resting on floor beams, which are assumed to act as nondeflecting supports. The only live-load response that need be calculated for the decking is transverse bending. If the decking is comprised of a concrete slab, it can be designed or evaluated by using the empirical method given in Sec. 6.2. Alternatively, live-load transverse moments can be obtained by the analytical method given in Sec. 6.4.

12.3   Analysis of Stringers

Since, as discussed above, the system of stringers and decking can be treated as a conceptual bridge supported on floor beams, the live-load longitudinal moments and shears in stringers can be obtained by the relevant D-type methods given in Chaps. 4 and 5.

While concrete decking in modern bridges is frequently made composite with the stringers by means of mechanical devices such as shear connectors, this is usually not the case with older bridges. Yet tests have shown that even in the absence of such devices, there exists an appreciable degree of composite action between the concrete slab and stringers (see Ref. 1). As a safe-side measure it is prudent to assume full composite action between the slab and stringers when calculating design longitudinal moments. As discussed in Chap. 2, the higher the flexural rigidity of a longitudinal member, the higher the share of live loads accepted by it. By analyzing the slab and stringers as fully composite, it is ensured that the live-load moments for stringers which result from the analysis will be greater than or equal to what the stringer will actually experience. On the other hand—a conservative approach should be maintained—if there are no mechanical shear connectors, the slab and stringers should be treated as noncomposite when resistances are calculated.

12.4   Analysis of Floor Beams

The mechanics of load distribution among floor beams can be readily visualized if the following simplified assumptions are made regarding the distribution of a concentrated load located between two adjacent floor beams.

1. The load is first dispersed by the decking and stringers, if present, to each of the two adjacent floor beams as distributed loads.

2. These two distributed loads are themselves partially transferred to other floor beams through the decking and stringers, if present, which relieves the directly loaded beams of some of the load.

This simplified model of load distribution is illustrated in Fig. 12.4.

Because of the absence of a simplified way of analyzing transfer of loads, it is usual to neglect the load transfer identified in step 2 above. It is also very tempting, and, in some cases, quite justifiable, to transfer a concentrated load to the two adjacent floor beams without accounting for the load dispersion by the deck slabs and stringers, identified in step 1 above. However, as shown in Ref. 2, such a simplification can lead to an overestimation of floor beam moments in excess of 30 percent in some cases.

The simplified method for analyzing live-load effects in floor beams given here consists of a two-part procedure corresponding to the two steps of load distribution identified above.

Load Dispersion by Decking and, if Present, Stringers

The longitudinal medium that is active in load dispersion comprises the decking and, if present, the stringers. The nature of the load dispersion is illustrated in Fig. 12.5. This shows the results of a grillage analysis [3] on an isolated panel of the floor system, subjected to a central concentrated

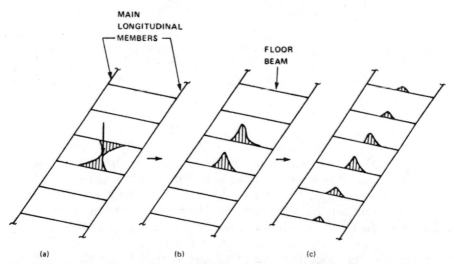

Figure 12.4 Mechanics of load distribution among floor beams: (a) concentrated load between two floor beams; (b) distributed load on adjacent floor beams; (c) transferred loads on other floor beams.

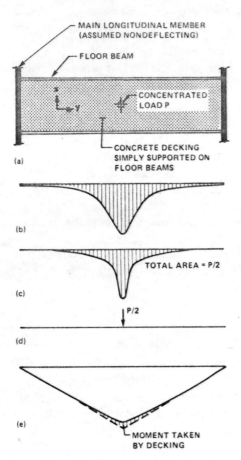

(a)

(b)

(c) TOTAL AREA = P/2

(d) P/2

(e) MOMENT TAKEN BY DECKING

Figure 12.5 Floor beam moments due to a concentrated load. (*a*) Partial plan of the floor system. (*b*) Longitudinal moment distribution in deck slab. (*c*) Distribution of "actual" longitudinal shear at the decking and floor beam interface. (*d*) "Assumed" load on a floor beam. (*e*) Diagram of floor-beam bending moment. Key: —Moment due to actual load; --Moment due to assumed point load.

load. The floor system consists of a concrete deck slab simply supported on two adjacent floor beams. The concentrated load is transferred to a floor beam through longitudinal shear at the interface of the beam and slab. As can be seen in Fig. 12.5c, the intensity of longitudinal shear is a maximum at a section directly opposite to the concentrated load, and diminishes rapidly as one moves away from the maximum intensity location. The floor beam bending moment diagram due to the interface longitudinal shear is shown in Fig. 12.5e, and is compared with the bending moment diagram due to a single concentrated load on the beam.

As can be seen, the difference between the two bending moment diagrams is not very large, and is limited to the maximum moment region. This difference between the two bending moment diagrams clearly represents a portion of bending moment which has to be sustained by the decking flexing about its own neutral axis. It can be concluded that any reduction of floor beam bending moment from that induced by the concentrated load can be effected only when the decking itself sustains some of the transverse moment. In practice, this reduction is significant only in floor systems which consist of a concrete deck slab supported directly by floor beams. For all other cases, the distribution of longitudinal shear is so highly localized that the actual floor beam moments are negligibly different from those caused by a single concentrated load. Hence, in such cases the original concentrated load can be replaced by statically equivalent concentrated loads on the two adjacent floor beams without significant error. The equivalent concentrated loads lie on the longitudinal line passing through the original concentrated load, and are obtained as the reactions of a simply supported beam carrying the concentrated load.

When the floor system consists of a concrete deck slab supported by floor beams, the statically equivalent concentrated loads on the floor beams, obtained as described above, can themselves be dispersed along the floor beam. This dispersion arises from the action of the deck slab and can be approximated by distributing the equivalent concentrated point load on the beam as a line load of triangular intensity, as shown in Fig. 12.6. The extent of the distributed load corresponds to a dispersion angle

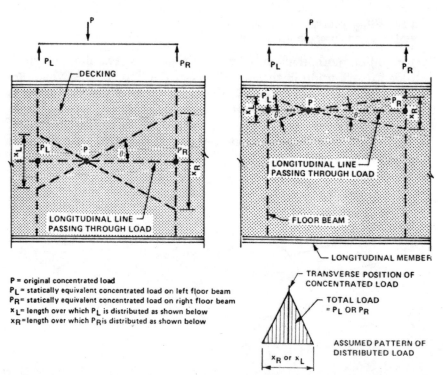

P = original concentrated load
P_L = statically equivalent concentrated load on left floor beam
P_R = statically equivalent concentrated load on right floor beam
x_L = length over which P_L is distributed as shown below
x_R = length over which P_R is distributed as shown below

TRANSVERSE POSITION OF CONCENTRATED LOAD

TOTAL LOAD = P_L OR P_R

ASSUMED PATTERN OF DISTRIBUTED LOAD

x_R or x_L

Figure 12.6 Dispersion of concentrated loads through concrete decking.

θ from the original concentrated load position of not more than 30°. As can be seen in Fig. 12.6, the extent of an equivalent distributed load near the longitudinal edge of the deck slab would be limited by the longitudinal edge of the decking.

Load Sharing among Floor Beams

If the longitudinal medium, i.e., the decking and stringers, is simply supported on floor beams, then clearly a load on a floor beam cannot be transferred to adjacent beams. However, if the longitudinal medium is continuous, then a load applied directly to a floor beam will be carried in part by the adjacent beams. The degree of load sharing depends upon the relative flexural and torsional rigidities of the longitudinal medium with respect to those of the floor beams, and also on the position of the load along the loaded floor beam. It is clear that a load located on the floor beam support will not be distributed to the adjacent beams; on the other hand, for a load well away from the support, distribution to adjacent beams will occur.

A Simplified Method for Load Distribution among Floor Beams

The simplified method given below is in two parts. The first part deals with load distribution in those portions of the floor beams which are situated transversely between the two main longitudinal members. In the case of pony truss and through-truss bridges, these portions constitute the entire floor system. The second part deals with those portions of the floor beams which are cantilevered out transversely from the main longitudinal members.

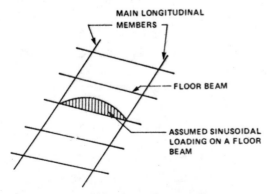

Figure 12.7 Assumed loading on a floor beam.

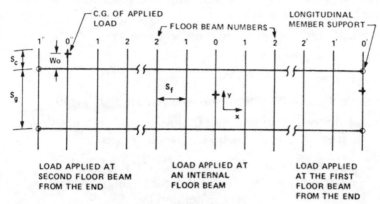

Figure 12.8   Floor beam numbering system.

Load Distribution among Floor Beams between Main Longitudinal Members

A large number of floor systems were analyzed by the grillage analogy method in order to develop the method given here. Preliminary analysis confirmed that the load distribution between floor beams is very little affected by the fact that the main longitudinal members themselves deflect under load. Hence, a key assumption in deriving the method is to treat the floor beams as being supported on rigid supports. Another key assumption is that the torsional rigidities of the floor system, being small, have a negligible effect on the load distribution among the floor beams.

Conceptually, any loading applied to a particular floor beam can first be analyzed harmonically into a series of loadings in the form of sine waves. Figure 12.7 shows the first such harmonic component applied to a floor beam. The method in its fullest form takes each harmonic of external load in turn and distributes it among the floor beams. In practice, it is found that only first harmonic effects are distributed in any appreciable degree, the higher harmonics of load being retained virtually entirely by the loaded floor beam. This is because the loadings accepted by the other floor beams arise directly from the deflection of the externally loaded one, and this deflection is dominated by the first harmonic. The following notation is employed:

F_i = the fraction of applied load taken by beam number i, where the beam number is as shown in Fig. 12.8 (the beam number depends upon the location of the loaded beam and the relative position of the beam under investigation with respect to the loaded beam)

D_x = longitudinal flexural rigidity per unit width of the longitudinal medium, which may consist of either decking or decking and stringers

D_y = transverse flexural rigidity per unit length as contributed by the floor beam

S_c = length of the cantilever portion of the floor beam

S_f = center-to-center spacing of floor beams

S_s = center-to-center spacing of stringers

S_g = center-to-center distance between main longitudinal members

W_0 = distance of the center of gravity of applied load on a cantilever floor beam from the nearest main longitudinal member

Some of the above notations are also illustrated in Fig. 12.8.

The pattern of load distribution between floor beams is then characterized by a dimensionless parameter ω defined by

$$\omega = \left(\frac{S_g}{S_f}\right)\left(\frac{D_x}{D_y}\right)^{0.25} \tag{12.1}$$

It will be noted that the parameter ω is similar to the flexural parameter θ which is used as a basis for the simplified methods of Chap. 4. Extensive analysis of many cases has confirmed the validity of ω as the characterizing parameter.

The Longitudinal Flexural Rigidity D_x

In calculating the value of ω, and in particular in establishing an appropriate value for D_x in the formula for ω, it is appropriate to consider how the longitudinal medium participates in load distribution.

In the floor systems of most real-life truss and similar bridges, a loaded floor beam is able to transfer significant loads only to its two nearest neighbors on either side. In other words, an externally applied load on one floor beam is eventually shared between five floor beams of which the externally loaded beam is the middle one. Hence, in order to take full advantage of load distribution, a loaded beam should have at least two beams on each side, and the longitudinal medium should provide flexural continuity over the five consecutive floor beams.

As discussed earlier, a longitudinal medium that is simply supported cannot transfer load from one floor beam to another. Hence, it is prudent to assume that those portions of the longitudinal medium which are not flexurally continuous over the floor beams do not participate in load transfer and, therefore, should not be included in the calculation of D_x. In the case of a longitudinal medium which consists of a concrete slab that is continuous over floor beams and steel stringers which are attached to steel floor beams only through web connections, the nature of the connections prevents the stringers from being fully continuous over the floor beams. In this case, although there is some degree of continuity of the stringers, it is safe and conservative to ignore their contributions, and to assume that the value of D_x is provided only by the deck slab.

Expressions for calculating D_x for various types of longitudinal mediums are given in Table 12.1. For those types which are not listed in the table, the value of D_x can be readily derived by keeping in mind the considerations mentioned above.

Transverse Flexural Rigidity D_y

The transverse flexural rigidity per unit length derives from the flexural rigidity of the floor beams, and is simply obtained by dividing the flexural rigidity EI of a floor beam by the floor beam spacing S_f.

TABLE 12.1 Expressions for Longitudinal Flexural Rigidity*

No.	Type of decking and stringers	Expression for D_x
1	Concrete slab without stringers but continuous over floor beams	$\dfrac{E_c t^3}{12}$
2	Concrete slab with stringers. Slab continuous over floor beams but stringers simply supported	$\dfrac{E_c t^3}{12}$
3	Concrete slab, noncomposite with stringers, both of which are continuous over floor beams	$\dfrac{E_c t^3}{12} + \dfrac{E_s I_s}{S_s}$
4	Concrete slab, composite with stringers, both of which are continuous over floor beams	$\dfrac{E_c}{S_s} I_{ss}$
5	Longitudinal nail-laminated wood deck continuous over floor beams	$\dfrac{E_L t^3}{12}$
6	Transversely post-tensioned, longitudinal laminated-wood deck, continuous over floor beams	$\dfrac{E_L t^3}{12}$
7	Transverse wood decking with stringers continuous over floor beams	$\dfrac{E_s I_s}{S_s}$

* Key: E_c = modulus of elasticity of concrete
E_L = longitudinal modulus of elasticity of wood
E_s = modulus of elasticity of stringer material
I_s = second moment of area of a stringer
I_{ss} = the combined second moment of area of slab and stringer, in concrete units
S_s = stringer spacing
t = decking thickness

In many timber truss bridges, a typical floor beam comprises a horizontal wood beam, a vertical wood post, and an inclined steel bar. The assembly, which is known as a *king-post truss*, is then as shown in Fig. 12.9. It can be readily appreciated that the equivalent flexural rigidity of a king-post truss varies along the span. However, for a central point load a beam of uniform second moment of area I, given by the following equa-

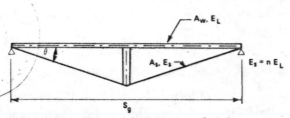

Figure 12.9 A king-post truss.

tion, has almost the same deflections along its length as the king-post truss.

$$I = \frac{S_g^2}{12}\left(\frac{\sin^2 \theta \cos \theta\, A_w A_s n}{\cos^3 \theta\, nA_s + A_w}\right) + I_w \tag{12.2}$$

where, as shown in Fig. 12.9, S_g is the spacing of main longitudinal components; A_s and A_w are the cross-sectional areas of the steel bar and wood beams, respectively; n is the ratio of the modulus of elasticity of steel to the longitudinal modulus of elasticity of wood; I_w is the second moment of area of the wood beam; and θ is the angle of inclination of the steel bar to the horizontal. The second moment of area of the equivalent beam, I, corresponds to the longitudinal modulus of elasticity of wood. Equation (12.2) is derived by neglecting the elastic shortening of the wood post.

While the equivalent second moment of area given by Eq. (12.2) is a good enough representation of the actual king-post truss subjected to a central point load, it is not so when the floor beam is subjected to loads away from the midspan. In such cases, the lack of analogy arises from the local bending of the wood beam. This local bending gives rise to substantial deflections and causes the maximum deflections to be moved away from the midspan even under concentrated loads. Obviously, a deflection response of that nature cannot be exactly replicated by a beam of uniform flexural rigidity. However, some compensation for the local bending can be made by reducing the flexural rigidity of the analogous beam so that under a uniformly distributed load its midspan deflection is equal to the sum of the corresponding midspan deflection of the actual floor beam and half the maximum local deflection of the wood beam. The revised second moment of area I_R of the beam which would satisfy the above deflection criterion can be obtained by the following expression.

$$I_R = \frac{I}{1 + 0.013m_c} \tag{12.3}$$

where I is obtained by Eq. (12.2), and $m_c = I/I_w$.

The effectiveness of I_R in closely reproducing the actual floor beam deflections is demonstrated in Fig. 12.10. This shows comparisons of actual deflections in the king-post truss with those of the analogous beam under various loading conditions. It can be seen that under a central concentrated load, which does not constitute a real-life loading case, the analogous beam deflections are substantially larger than the actual ones along the whole span. However, under a uniformly distributed load, the beam deflections are larger than the truss deflections at midspan, but smaller at points away from the midspan. The beam deflections follow the irregular deflection profile of the truss as closely as possible. Such is again the case where the floor beam is loaded by four concentrated loads. It is noted that this loading case represents the real-life loading due to two design vehicles.

Rigorous analyses have confirmed that, for the purpose of the simplified analysis given below, the representation of a king-post truss by a beam of uniform second moment of area I_R is quite satisfactory.

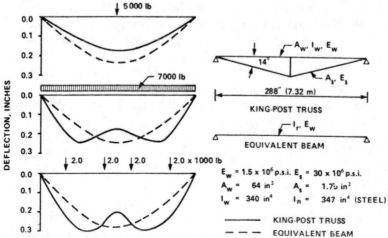

Figure 12.10 Equivalent beam deflections versus actual king-post truss deflections.

Load Distribution among Floor Beams
Transversely Contained between
Longitudinal Members

In accordance with the method described earlier for load dispersion by the longitudinal medium, all concentrated loads are transferred to the immediately adjacent floor beams. Such loads are then themselves transferred to other floor beams, in accordance with the simplified method, which involves the following calculation steps.

1. Calculate the effective longitudinal flexural rigidity D_x of the longitudinal medium, using the expressions of Table 12.1.

2. Calculate the equivalent transverse flexural rigidity D_y of a floor beam in the manner described earlier. In a king-post truss obtain I_R, using Eqs. (12.2) and (12.3), and then obtain D_y as $E_L I_R / S_f$.

3. Calculate the value of ω, using Eq. (12.1).

4. Corresponding to the value of ω, obtain values of distribution factors F_0, F_1, and F_2 from Fig. 12.11. These distribution factors are to be used for floor beams between the two main longitudinal members.

5. Multiply the load applied to floor beam 0 by the factors F_0, F_1, and F_2 so as to give reduced loads for the calculation of moments and shears in floor beams 0, 1, and 2, respectively.

$F_0 + 2F_1 + 2F_2$ frequently may exceed 1.0. This happens when floor beams farther away from beam 0 experience uplift. The reduction of net loads due to such uplift is, however, neglected, so that the method has a degree of conservatism within it. The procedure described above is applicable only if there are at least two floor beams on either side of the loaded beam and the longitudinal medium provides flexural continuity over all five floor beams. For other cases, F_0, F_1, and F_2 should be obtained as shown above, but with the following modification.

When the loaded floor beam is the first floor beam from the end, it is numbered 0′, as shown in Fig. 12.8, and the adjacent beams are numbered as 1′ and 2′. Load fractions F_0', F_1', and F_2' can be approximately obtained as follows:

$$F_0' = F_0 + F_1 + F_2$$
$$F_1' = F_1 + F_2 \qquad\qquad (12.4)$$
$$F_2' = F_2$$

When the loaded floor beam is the second one from the end, it is identified by number 0″ as shown in Fig. 12.8; the first beam from the end is identified by 1″ and the beams on the other side of the loaded beam by 1 and 2. In this case, load fractions for beam numbers 1 and 2 are the same as obtained from Fig. 12.11. However, load fractions F_0'' and F_1'' are slightly increased according to the following approximate relationships:

$$F_0'' = F_0 + F_2$$
$$F_1'' = F_1 + F_2 \qquad (12.5)$$

The above relationships could also be applied when the longitudinal medium becomes flexurally discontinuous at the second floor beam from the loaded one. When a loaded floor beam is at a flexural discontinuity in the longitudinal medium, the load fraction F_0 should be increased by an amount equal to F_2.

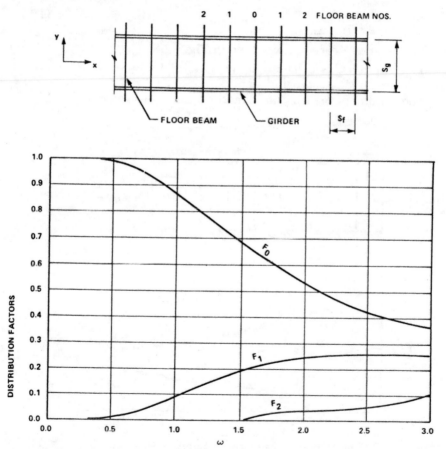

Figure 12.11 Distribution factors for floor beams.

Method for Load Distribution among Cantilever Floor Beams

The method for analyzing load distribution effects between those floor beams which are cantilevered transversely beyond the main longitudinal members also makes use of the characterizing parameter ω. However, the formula for this parameter is now modified in order to take account of the position of the center of gravity of the load on the floor beam with respect to the cantilever support. The redefined parameter is obtained by the following equation:

$$\omega = \frac{2S_e}{S_f}\left(\frac{D_x}{D_y}\right)^{0.25}\left[\frac{\overline{m}^2(4\overline{m}-3)}{0.625}\right]^{0.25} \tag{12.6}$$

where

$$\overline{m} = \frac{W_0}{2S_g} \tag{12.7}$$

and S_g, S_f, and W_0 are as illustrated in Fig. 12.8.

The sequence of calculation steps is the same as for floor beams between main longitudinal girders.

The validity of the simplified method is demonstrated in Fig. 12.12,

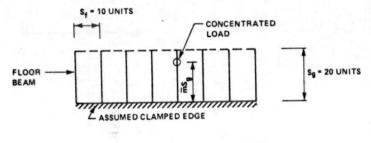

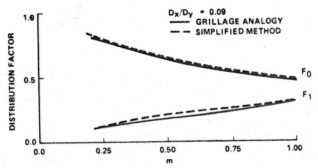

Figure 12.12 Comparison of distribution factors obtained by the simplified and grillage methods.

in which values of F_0 and F_1 obtained by this method for a particular flooring system are compared with those obtained by the computer-based grillage analogy method. It can be seen that the agreement between the two methods is very good.

12.5 Example of a Floor System Incorporating King-Post Trusses

There are so many varieties of floor systems in truss and similar bridges that it is not practicable to deal explicitly with all possible types in a single chapter. However, if the principles underlying the methods of analysis are fully comprehended, then it should not be difficult for an engineer to apply them to other floor systems not explicitly dealt with here. A numerical example is provided here not only to illustrate the use of the method but also in the belief that in following through a completely worked problem numerically some of the basic principles involved will become more fully understood.

In this example, a floor system which is commonly encountered in timber truss bridges is analyzed. The floor system consists of king-post trusses spaced at 7-ft 6-in (2.29-m) centers and spanning between main longitudinal trusses 28 ft (8.53 m) apart, wood stringers spanning between the king-post trusses, and a transverse nail-laminated wood decking. Details of the king-post trusses, along with some other relevant details of the floor system, are given in Fig. 12.13. The stringers have a cross section of $4\frac{1}{2} \times 11\frac{1}{4}$ in (114×292 mm), are spaced at 18 in (457 mm) center to center, and are continuous over three adjacent king-post trusses. The decking consists of 2×6 in (51×152 mm) transverse laminates nailed together. The longitudinal medium is thus continuous over only two panels. It should be noted that at sections of flexural discontinuity the stringers project about 6 in (150 mm) on either side of a floor beam. The decking prevents these projections from lifting, so that some flexural continuity is provided even at these sections.

The floor system is analyzed under AASHTO HS 20 loading. The calculations which follow do not include any analysis of timber decking; this component of the floor system is usually designed by empirical methods.

Analysis of Stringers

Since the stringers are continuous over two panels, the conceptual "local bridge" is double-span; details of this conceptual bridge are shown in Fig. 12.14. The figure also shows the effective span lengths for negative

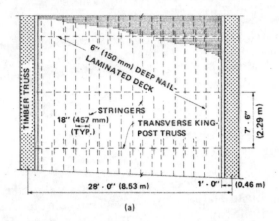

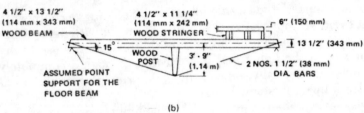

Figure 12.13 Details of a floor system incorporating king-post trusses as floor beams.

and positive moment regions which have been calculated according to the guidance given in Fig. 8.8. The relevant details of the "bridge" are as follows.

Effective span for negative moment regions = 3.0 ft (0.91 m)
Effective span for positive moment regions = 6.0 ft (1.83 m)
Width $2b$ = 25.0 ft (7.62 m)
Curb-to-curb width = 25.0 ft (7.62 m)
Number of design lanes = 2
Design lane width = 12.5 ft (3.81 m)

From Eq. (2.13), G_{LT} is taken to be equal to $0.07E_L$. Using Eqs. (2.14), the plate rigidities are calculated as follows:

$$D_x = E_L \frac{4.5 \times 11.25^3}{12(18)} = 29.7E_L$$

$$D_y = E_L \frac{6.0^3}{12} = 18.0E_L$$

The value of K for w/d of $11.25/4.5$ (= 2.5) is found from Fig. 2.6 to be 0.25.

$$D_{xy} = \frac{0.07E_L(11.25)(0.25 \times 4.5^3)}{18} = 1.00E_L$$

$$D_{yx} \qquad\qquad\qquad\qquad = 0.0$$

$$D_1 = D_2 \qquad\qquad\qquad = 0.0$$

Thence, using Eqs. (4.4),

$$\alpha = \frac{1.00}{2(29.7 \times 18.0)^{0.5}} = 0.02$$

For the negative moment regions, using Eq. (4.5), one finds

$$\theta = \frac{25.0}{2(3.0)}\left(\frac{29.7}{18.0}\right)^{0.25} = 4.72$$

while for the positive moment regions, again using Eq. (4.5),

$$\theta = \frac{25.0}{2(6.0)}\left(\frac{29.7}{18.0}\right)^{0.25} = 2.36$$

It is noted that the value of θ for negative moment regions is outside the range for which D values are given in Fig. 4.19. In this case, as advised in Sec. 8.4, the θ value is taken to be 2.5. Using Eq. (4.9),

$$\mu = \frac{12.5 - 11.0}{2} = 0.75$$

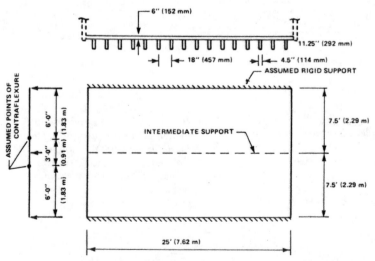

Figure 12.14   Conceptual local bridge for the analysis of stringers.

From the charts given in Fig. 4.19a, and putting $\theta = 2.5$ and $\alpha = 0.02$ for the negative moment regions, one finds

D for external portions $= 1.64$ m

D for internal portions $= 1.30$ m

C_f $\qquad\qquad = 2.0\%$

Thence, using Eq. (4.10), the following D_d values are obtained:

$$D_d \text{ for external portions} = 1.64\left[1 + \frac{0.75(2)}{100}\right] = 1.66 \text{ m (5.45 ft)}$$

$$D_d \text{ for internal portions} = 1.30\left[1 + \frac{0.75(2)}{100}\right] = 1.32 \text{ m (4.33 ft)}$$

The maximum negative moment due to one HS 20 vehicle is 23,000 lb · ft (31.2 kN · m), as shown in Fig. 12.15. Therefore, the maximum negative moment for each external stringer is found to be equal to (1.5/5.45)(23,000/2), or 3165 lb · ft (4.3 kN · m). It is noted that, because of the narrow spacing of the stringers, each of the three outermost stringers should be regarded as an external one.

Similarly, the maximum negative moment for each internal stringer is equal to (1.5/4.33)(23,000/2), or 3983 lb · ft (5.4 kN · m).

The maximum positive moment in each external stringer is found to

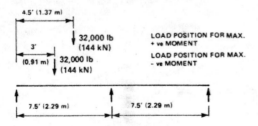

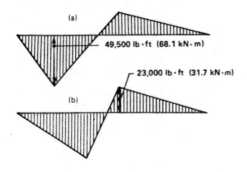

Figure 12.15 Bending moments in stringers due to one HS 20 vehicle. (*a*) Bending moment diagram corresponding to maximum positive moments. (*b*) Bending moment diagram corresponding to maximum negative moments.

be 6812 lb · ft (9.2 kN · m), and the maximum positive moment in each internal stringer is found to be 8574 lb · ft (11.6 kN · m).

The D value for longitudinal shear is found with reference to Table 5.3 and Eq. (5.11).

$$D = 5.25 \left(\frac{1.5}{6.56} \right)^{0.25} = 3.63 \text{ ft } (1.11 \text{ m})$$

The maximum shear due to one HS 10 vehicle is very nearly equal to 32,000 lb (144 kN) for the conceptual local bridge under consideration. Therefore, longitudinal shear per stringer is equal to (1.5/3.63) (32,000/2), or 6611 lb (29.0 kN).

Analysis of Floor Beams

The ratio of the modulus of elasticity of steel to the longitudinal modulus of wood, that is, n, is taken to be 20. Using Eq. (12.2), I is calculated as follows:

$$I = \frac{(28.0 \times 12)^2}{12} \left[\frac{(\sin^2 15°)(\cos 15°)(13.5 \times 4.5)(2\pi)(0.25)(1.5^2)(20)}{(\cos^3 15°)(20)(2\pi)(0.25)(1.5^2) + (13.5 \times 4.5)} \right]$$

$$+ \frac{4.5 \times 13.5^3}{12}$$

$$= 21,928 \text{ in}^4 (0.0091 \text{ m}^4)$$

From Eq. (12.3),

$$I_R = \frac{21,928}{1 + 0.013(21,928/922)} = 16,749 \text{ in}^4 (0.0070 \text{ m}^4)$$

from which

$$D_y = \frac{16,749}{7.5 \times 12} E_L = 186.1 E_L$$

Ignoring the contribution of the transverse laminated-wood decking, D_z is obtained as follows:

$$D_z = \frac{4.5 \times 11.25^3}{12 \times 18} E_L = 29.7 E_L$$

From Eq. (12.1),

$$\omega = \frac{28 \times 12}{7.5 \times 12} \left(\frac{29.7 E_L}{186.1 E_L} \right)^{0.25} = 2.36$$

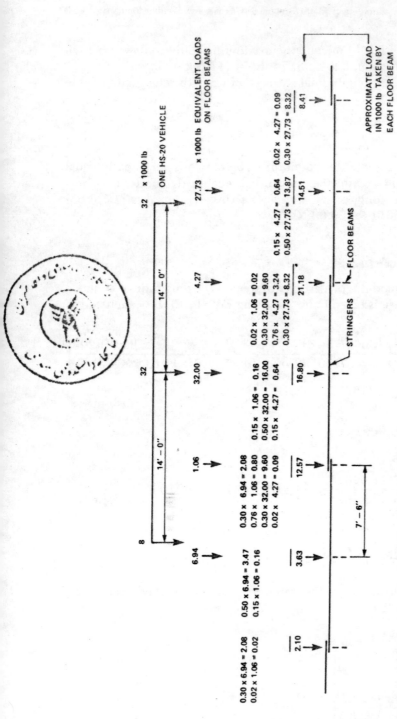

Figure 12.16 Apportioning of loads to floor beams.

From Fig. 12.11, with ω equal to 2.36, F_0, F_1, and F_2 are found to be equal to 0.46, 0.26, and 0.04, respectively. For a load on the floor beam over which the stringers are continuous, Eq. (12.5) should be used to account for the effect of discontinuity of the longitudinal medium beyond the second floor beam from the loaded one.

$$F_0'' = 0.46 + 0.04 = 0.50$$

$$F_1'' = 0.26 + 0.04 = 0.30$$

When the load is over the floor beam under the stringer discontinuity, the situation shown in Fig. 12.8 for "end-loaded" behavior is obtained, except that now such behavior is applied on both sides of the position 0'; in other words the pattern of loaded girders becomes 2', 1', 0', 1', 2'. Equations (12.4) are used, with the modification that the F_1' value is now divided between two beams, as is the F_2' value.

Using Eqs. (12.4), we have

$$F_0' = 0.46 + 0.26 + 0.04 = 0.76$$

$$F_1' = 0.26 + 0.04 \qquad\quad = 0.30$$

$$F_2' = 0.04$$

Hence the load fractions taken by the five beams are 0.02, 0.15, 0.76, 0.15, and 0.02.

By using the various coefficient values, loads due to a single HS 20 vehicle are apportioned to the various floor beams as shown in the calculations given in Fig. 12.16.

It is noted that these calculations correspond to only one longitudinal vehicle position. Different equivalent loads on each floor beam will be obtained for a different longitudinal vehicle position.

References

1. Bakht, B., and Csagoly, P. F.: *Bridge Testing*, Structural Research Report 79-SRR-10, Ministry of Transportation and Communications, Downsview, Ontario, Canada, 1979.
2. Bakht, B., and Csagoly, P. F.: Load carrying capacity of highway bridges, *3d IABSE Conference on Structural Safety and Reliability*, Trondheim, Norway, 1981.
3. Jaeger, L. G., and Bakht, B.: The grillage analogy in bridge analysis, *Canadian Journal of Civil Engineering*, 9(2), 1982, pp. 224–235.

13

METHODS FOR ANALYZING DECK SLAB OVERHANGS

13.1 Introduction

Transverse moments due to live loading consisting of one or more concentrated loads, in concrete deck slab overhangs of slab-on-girder and other similar types of structures, are investigated in this chapter. The portions of cross sections of deck slabs which are here referred to as *overhangs* are shown shaded in Fig. 13.1. As discussed in Sec. 3.5, these components are treated as cantilever slabs on unyielding supports. This assumption, which is employed even in refined analyses, leads to negligible errors, as shown in Ref. 5.

Longitudinal and Transverse Directions

The nomenclature of directions in a cantilever slab overhang may cause some confusion. In the case of a cantilever slab in isolation, the term *length* usually refers to the distance between the root of the cantilever and the free parallel edge, so that the transverse direction becomes the one which is parallel to the root. However, in the nomenclature of the whole bridge, the longitudinal direction is the direction of traffic flow, and this is parallel to a girder and hence to the root of the cantilever. In this book we adhere to the nomenclature for the whole bridge. Accordingly, the length of the overhang refers to the direction parallel to the root.

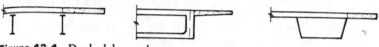

Figure 13.1 Deck slab overhangs.

Assumption of Isotropy

The simplified methods given in this chapter were developed by assuming the slab material, i.e., the reinforced or prestressed concrete, to be homogeneous and isotropic. This assumption leads to some overestimation of the transverse moments for the following reasons.

Transverse moments due to concentrated loads in cantilever slabs are larger than the corresponding moments in the longitudinal direction. This results in more longitudinally oriented cracks than transverse ones. It is well known that cracking of concrete tends to reduce the flexural rigidity of the component in the plane perpendicular to the crack. Longitudinal cracks reduce the transverse flexural rigidity of the cantilever slab, while leaving the rigidity in the longitudinal direction more or less unaffected.

The reduction of flexural rigidity in the transverse direction due to longitudinal cracks is diagrammatically shown in Fig. 13.2, in which the longitudinal cracks are replaced by grooves. With the help of this figure, it can be readily appreciated that the higher flexural rigidity of the slab in the longitudinal direction helps to distribute the load more effectively than would be the case if the rigidities in the two directions were equal.

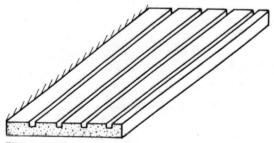

Figure 13.2 Diagrammatic representation of longitudinal cracks.

Unfortunately, there does not exist today any simple method of analysis, nor even any readily available refined method of analysis, which takes account of the effect discussed above. Until such methods are available, one must rely on analyses which treat the cantilever slab as being composed of isotropic material. The simplified methods given in this chapter were developed by using the results of refined linear elastic analyses. The errors arising from assumptions implicit in these refined methods are, therefore, also implicit in the simplified methods. The simplified methods are applicable, sufficiently accurately for design purposes, to cantilever slabs of linearly varying thickness.

13.2   Method for Unstiffened Infinite Cantilever Slabs

The effect of a concentrated load on a cantilever slab becomes insignificant beyond a distance of about $2a$ in the longitudinal direction of the slab, where a is the cantilever slab width. Therefore, the middle portions of a cantilever slab overhang remote from both ends (as identified in Fig. 13.3) can be regarded and analyzed as portions of a slab of infinite length. Portions near a free transverse edge can be regarded as part of a semi-infinite cantilever slab, i.e., of a slab which extends to infinity in one direction. Analysis of these latter portions is dealt with in Sec. 13.3.

Transverse moments per unit length, M_{yc}, due to a concentrated load P on an infinitely long cantilever slab with linearly varying thickness in the transverse direction, are given by:

$$M_{yc} = \frac{P}{\pi} A' \left[\frac{1}{\cosh A'x/(c - y)} \right] \tag{13.1}$$

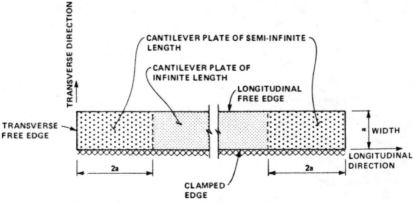

Figure 13.3   Nomenclature used for cantilever slabs.

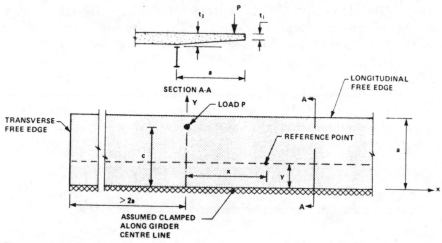

Figure 13.4   Deck slab overhang as an infinitely wide cantilever plate.

where the notation is as shown in Fig. 13.4, and A' is a coefficient which depends upon the load and reference point locations, and on the thickness ratio of the slab, which is the ratio of the thickness at the root to that of the longitudinal free edge of the slab. Values of A' can be read directly from the charts given in Fig. 13.5 for thickness ratios of 1, 2, and 3, and those for intermediate values can be obtained by linear interpolation.

It should be noted that Eq. (13.1) is valid only when y is smaller than c.

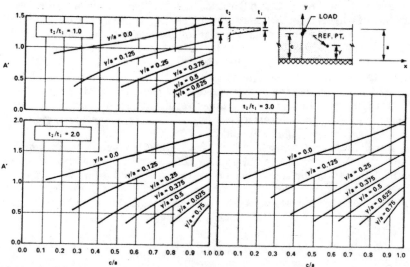

Figure 13.5   Charts for A' for unstiffened cantilever slabs.

It gives the value of M_{yc} at the reference point for a load at the position shown. The method which led to Eq. (13.1) is developed in Refs. 3 and 4.

An alternative form of Eq. (13.1) may be preferred. Setting $A'x/(c - y)$ equal to k, we have, from the definition of the cosh function,

$$M_{yc} = \frac{2P}{\pi} A' \left(\frac{1}{e^k + e^{-k}} \right) \tag{13.2}$$

The moment intensity M_{yc} is maximized when the concentrated load is as far away from the root as possible. Therefore, for obtaining the govern-

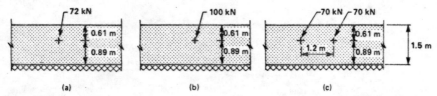

(a)   (b)   (c)

Figure 13.6 Cantilever slab of infinite length under AASHTO and Ontario loadings. (a) AASHTO HS 20 loading. (b) Ontario loading, single-axle. (c) Ontario loading, dual-axle.

ing live-load cantilever transverse moments, the heaviest axle (or pair of axles) should be placed as close to the longitudinal free edge as is permitted by the relevant code of practice.

Example 13.1 Let us find the live-load cantilever transverse moment intensities due to both the AASHTO HS 20 and Ontario loadings (Figs. 4.16 and 4.6, respectively), at the cantilever root in a 1.5-m-wide (4.92-ft-wide) cantilever slab of constant thickness and in a tapered slab having a thickness ratio of 2.0.

For both the AASHTO and Ontario loadings, the distance of the center of the line of wheels from the longitudinal free edge is taken to be 0.61 m (2 ft). Clearly, the cantilever overhang can accommodate only one line of wheels. For the AASHTO HS 20 loading, the axles are 4.27 m (14 ft) apart. The influence of a load which is longitudinally at a distance of 4.27 m (14 ft), that is, $2.84a$, from the reference point is negligible. Therefore, wheels of the HS 20 vehicle other than the immediate one are excluded from consideration. For the Ontario vehicle, however, the governing load position is not immediately obvious, and it is necessary to find moments for the heaviest single and dual axles shown in Fig. 13.6b and c.

For all load cases shown in Figs. 13.6a, b, and c, y/a is equal to zero and c/a is equal to 0.59. For the slab of uniform thickness, i.e., one with a thickness ratio of 1.0, A' for the above y/a and c/a values is found from Fig. 13.5 to be 1.17; for the slab with a thickness ratio of 2.0, A' is 1.47. For the former slab the maximum cantilever transverse moments due to the three loading cases identified in Fig. 13.6 are calculated as follows:

$$(a) \quad M_{yc} = \frac{72}{\pi} (1.17) \left[\frac{1}{\cosh{(1.17 \times 0.0)/0.89}} \right]$$

$$= 26.8 \text{ kN} \cdot \text{m/m} \ (6025 \text{ lb} \cdot \text{ft/ft})$$

(b) $M_{yc} = \dfrac{100}{\pi}(1.17)$

   $= 37.2 \text{ kN} \cdot \text{m/m } (8363 \text{ lb} \cdot \text{ft/ft})$

(c) $M_{yc} = 2\left(\dfrac{70}{\pi}\right)(1.17)\left[\dfrac{1}{\cosh\,(1.17 \times 0.6)/0.89}\right]$

   $= 39.5 \text{ kN} \cdot \text{m/m } (8880 \text{ lb} \cdot \text{ft/ft})$

Hence the governing load case for the Ontario loading is case (c) with a dual-axle load. It is noted that in this case the maximum value of M_{yc} is obtained at the point on the root of the cantilever which is equidistant from the two loads.

As noted above, for the slab with a thickness ratio of 2.0, the value of A' is found from Fig. 13.5 to be 1.47, and thence the maximum transverse cantilever moments for the three load cases shown in Fig. 13.6 are calculated as follows:

(a) $M_{yc} = \dfrac{72}{\pi}(1.47)$

   $= 33.7 \text{ kN} \cdot \text{m/m } (7576 \text{ lb} \cdot \text{ft/ft})$

(b) $M_{yc} = \dfrac{100}{\pi}(1.47)$

   $= 46.8 \text{ kN} \cdot \text{m/m } (10{,}521 \text{ lb} \cdot \text{ft/ft})$

(c) $M_{yc} = 2\left(\dfrac{70}{\pi}\right)(1.47)\left[\dfrac{1}{\cosh\,(1.47 \times 0.6)/0.89}\right]$

   $= 42.7 \text{ kN} \cdot \text{m/m } (9599 \text{ lb} \cdot \text{ft/ft})$

It is noted that the increase in thickness ratio has changed the governing load case for Ontario loading. In the slab with uniform thickness the maximum intensity of moment, as shown earlier, is governed by the two-axle loading, and in the case under consideration, by the one-axle loading.

The above calculations do not include allowances due to impact or dynamic loads.

13.3 Method for Unstiffened Semi-infinite Cantilever Slabs

This method is applicable to portions in the vicinity of transverse free edges of cantilever deck slab overhangs, as identified in Fig. 13.3. According to this method, which is developed in Ref. 2, the moments M_{yc} at the root of a cantilever slab due to a concentrated load in the vicinity of a transverse free edge are given by:

$$M_{yc} = \frac{PA'}{\pi}\left[\frac{1}{\cosh\,(A'x/c)} + B'e^{-Kx/a}\right] \tag{13.3}$$

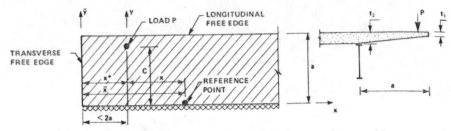

Figure 13.7   Deck slab overhang as cantilever slab of semi-infinite width.

where the notation is as shown in Fig. 13.7; A' is obtained from Fig. 13.5 for y/a equal to 0.0; B' is obtained from Fig. 13.8; and K is calculated from the following expression:

$$K = \frac{a}{c} \frac{A'B'}{2} \cdot \frac{1}{\tan^{-1} e^{-A\pi x/c}} \tag{13.4}$$

The above method, unlike the one described in Sec. 13.2, is available only for reference points along the cantilever root.

Example 13.2   The use of the above method is illustrated in this example, in which a 1.5-m-wide (4.92-ft-wide) cantilever slab of uniform thickness is analyzed for the three load cases shown in Fig. 13.9. Load case *(a)* involves a wheel of the AASHTO HS 20 vehicle, while cases *(b)* and *(c)* correspond to the single and dual axles of the Ontario loading, respectively.

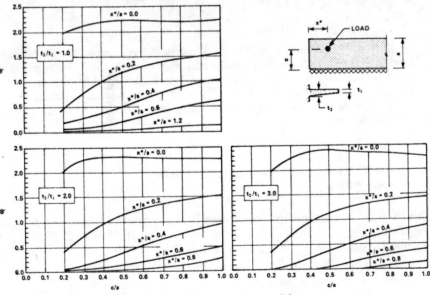

Figure 13.8   Charts for B' for unstiffened cantilever slabs.

For load case *(a)* the quantities c/a, y/a, and x^*/a are equal to 0.59, 0.0, and 0.0, respectively. The corresponding values of A' and B' (from Figs. 13.5 and 13.8) are 1.17 and 2.2. In seeking the intensity of the cantilever transverse moment at the transverse free edge, it is noted that $\bar{x}$ is equal to 0.0 at this point, and that consequently the second term within the brackets of Eq. (13.3) reduces to B', so that there is no need to obtain the value of K. In this case the maximum value of M_{yc} is given by

$$M_{yc} = \frac{72(1.17)}{\pi}(1 + 2.20) = 85.8 \text{ kN} \cdot \text{m/m (19,288 lb} \cdot \text{ft/ft)}$$

For $\bar{x}$ and x^* both equal to zero, the first term within the brackets corresponds to M_{yc} in an infinitely long cantilever plate and the second term repre-

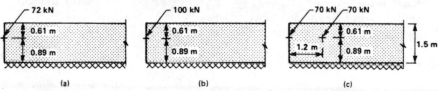

(a)   (b)   (c)

Figure 13.9 Cantilever slab of semi-infinite length under AASHTO and Ontario loadings. *(a)* AASHTO HS 20 loading. *(b)* Ontario loading, single-axle. *(c)* Ontario loading, dual-axle.

sents the increase in M_{yc} due to the vicinity of the transverse free edge. It can be seen that the load on the transverse free edge induces a moment intensity which is 220 percent larger than the corresponding moment in the middle portions of the overhang. However, as can be seen from what follows, the increased size of M_{yc} is of a fairly localized nature.

For load case *(b)* (see Fig. 13.9), the maximum moment can be obtained simply by prorating that for load case *(a)*. Hence the maximum moment is $100/72 \times 85.8$, or 119.2 kN $\cdot$ m/m (26,796 lb $\cdot$ ft/ft).

Still with load case *(b)* it is instructive to examine M_{yc} at the root of the cantilever at a distance of 0.5 m (1.64 ft) from the transverse free edge. Hence $\bar{x}$ is equal to 0.5 m (1.64 ft). Substituting the values of c/a, A', and B' in Eq. (13.4) yields

$$K = \frac{1}{0.59}\frac{1.17(2.2)}{2}\left(\frac{1}{\tan^{-1} e^{0.0}}\right) = 2.78$$

It is noted that the expression $\tan^{-1} e^{0.0}$ is the angle whose tangent is 1.0 since $e^{0.0}$ is equal to unity, and that this angle must always be expressed in radians, being $\pi/4$ rad in the present case.

Since the load is on the transverse free edge, both x and $\bar{x}$ are equal to 0.5 m (1.64 ft). Hence from Eq. (13.3),

$$M_{yc} = \frac{100(1.17)}{\pi}\left[\frac{1}{\cosh(1.17 \times 0.5)/0.89} + 2.2e^{-2.78(0.5)/1.5}\right]$$
$$= 63.30 \text{ kN} \cdot \text{m/m (14,230 lb} \cdot \text{ft/ft)}$$

At a distance of 1.0 m (3.28 ft) from the transverse free edge M_{yc} is found to be 31.3 kN · m/m (7036 lb · ft/ft), and at a distance of 1.50 m (4.92 ft), it is equal to 15.3 kN · m/m (3439 lb · ft/ft). Thus, it can be seen that within a distance of a, the intensity of moment is reduced from 119.2 to 15.3 kN · m/ m (26,796 to 3439 lb · ft/ft), a reduction of 87 percent.

Recognizing the highly localized nature of the peak intensity of M_{yc} in the vicinity of the transverse free edge, the Ontario code does not require its calculation; instead it requires that for a distance a from an unsupported transverse edge, the amount of reinforcement provided shall be at least twice that required in the interior portions of the deck slab overhang.

It remains to calculate M_{yc} at the transverse free edge due to load case (c), shown in Fig. 13.9.

From Figs. 13.5 and 13.8, A' and B' for the load on the transverse free edge are 1.17 and 2.2, respectively; for the second load, A' remains the same but B' is found to be 0.23 from Fig. 13.8 for $x*/a$ equal to 0.8. In neither case is the value of K required, since $\bar{x}$ is equal to 0.0.

M_{yc} due to the two loads is given by

$$M_{yc} = \frac{70(1.17)}{\pi}(1 + 2.20)$$

$$+ \frac{70(1.17)}{\pi}\left[\frac{1}{\cosh (1.17 \times -1.2)/1.5} + 0.23\right]$$

$$= 83.4 + 23.8$$

$$= 107.2 \text{ kN} \cdot \text{m/m} \ (24{,}098 \text{ lb} \cdot \text{ft/ft})$$

The above moment is larger than that which occurs at the position on the root of the cantilever midway between the two point loads, because of the "concentrating" effect of the transverse free edge. For the same reason the moment due to the heavy single load [calculated in case (b) as 119.2 kN · m/m] is larger than that caused by the wheels of the dual axle.

13.4 Method for Edge-Stiffened Infinite Cantilever Slabs

Most deck slab overhangs have appendages to stiffen their longitudinal free edges. These appendages may take the form of small curbs or heavy barrier walls which are monolithic with the deck slab, as shown in Fig. 13.10. Edge beams of the type shown in Fig. 13.10 can enhance load dispersion considerably.

"Concealed beams" comprising increased reinforcement within the depth of the slab do not appreciably increase the flexural rigidity of the edge, and hence are not counted as edge beams for the method of analysis given here.

As discussed in Ref. 1 the method of analyzing edge-stiffened cantilever slabs of infinite length is almost the same as that for corresponding
unstiffened slabs, except that values of the coefficient A' are obtained
from a different set of charts. The value of A' now also depends upon the

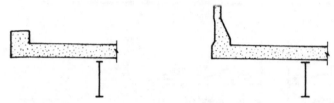

Figure 13.10 Examples of edge stiffening of cantilever slab
overhangs.

ratio of the flexural rigidity of the edge beam to the total flexural rigidity
of the cantilever slab about its own neutral axis. Demarcation of the
cantilever slab and edge beam is shown in Fig. 13.11, together with other
relevant notation.

Transverse cantilever moments due to a concentrated load are obtained by using Eq. (13.1), where the value of coefficient A' is obtained
from the relevant chart of Fig. 13.12. As can be seen in the figure, A'
depends upon the thickness ratio of the slab, the ratio of edge beam and
cantilever slab rigidities, and the positions of loads and reference points.
The method is applicable only when y is smaller than c.

For a slab of linearly varying thickness from t_2 to t_1, and which is of
width a, the second moment of area I_s is given by

$$I_s = \frac{a}{48}(t_2^3 + t_2^2 t_1 + t_2 t_1^2 + t_1^3) = \frac{a}{48} \frac{t_2^4 - t_1^4}{t_2 - t_1} \qquad (13.5)$$

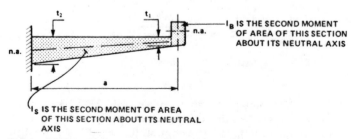

I_B IS THE SECOND MOMENT
OF AREA OF THIS SECTION
ABOUT ITS NEUTRAL AXIS

I_S IS THE SECOND MOMENT OF AREA
OF THIS SECTION ABOUT ITS NEUTRAL
AXIS

Figure 13.11 Cross section of an edge-stiffened cantilever slab.

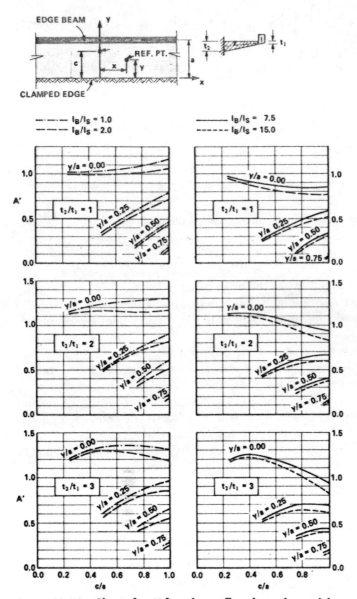

Figure 13.12   Charts for A' for edge-stiffened cantilever slabs.

Example 13.3   Figure 13.13 shows an example of a cantilever slab with stiffened longitudinal edge. This is the same slab which was analyzed in Sec. 13.2 for AASHTO loading (Fig. 13.6a), with the edge stiffening added. From the slab and edge beam dimensions shown in Fig. 13.13, the second moments of area I_B and I_S are calculated as follows:

$$I_B = \frac{0.2 \times 0.46^3}{12} = 1.62 \times 10^{-3} \text{ m}^4$$

$$I_S = \frac{1.4 \times 0.22^3}{12} = 1.24 \times 10^{-3} \text{ m}^4$$

Hence I_B/I_S is equal to 1.31.

The maximum intensity of cantilever transverse moment, M_{yc}, occurs at the root of the cantilever slab, i.e., with $x = y = 0$. Hence y/a is equal to 0.0 and c/a is equal to 0.59. Referring to Fig. 13.12 and to the chart for which

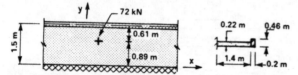

Figure 13.13 Edge-stiffened cantilever slab of infinite length under AASHTO wheel load.

$t_2/t_1 = 1.0$, one finds that the values of A' corresponding to $I_B/I_S = 1.0$ and $I_B/I_S = 2.0$ are 1.04 and 1.00, respectively. By linear interpolation the value of A' corresponding to $I_B/I_S = 1.31$ is equal to 1.03. Hence, by Eq. (13.1), the maximum intensity of transverse cantilever moment, M_{yc}, is given by

$$M_{yc} = \frac{72(1.03)}{\pi} = 23.6 \text{ kN} \cdot \text{m/m (5305 lb} \cdot \text{ft/ft)}$$

It is interesting to recall that the corresponding value of M_{yc} without the edge beam is 26.8 kN · m/m (6025 lb · ft/ft), so that the inclusion of the rather small edge beam has resulted in a reduction in the maximum value of M_{yc} of about 12 percent.

13.5 Method for Edge-Stiffened Semi-infinite Cantilever Slab

The reduction in intensity of M_{yc} referred to in Sec. 13.4 arises from the fact that the edge stiffening helps to distribute the effects of the concentrated load more effectively in the longitudinal direction.

In the case of an unstiffened slab it has already been noted in Sec. 13.3 that the moments at the transverse free edge are highly localized. This leads to a design approach, usually sufficiently accurate, of regarding an outside length a of the slab, measured from the transverse free edge, as requiring the "semi-infinite" treatment, while the remainder of the slab is treated as "infinite." This is the approach taken by the Ontario code. However, when the slab is edge-stiffened longitudinally, the load effects are more effectively distributed in the longitudinal direction, and it is recommended that an outside length of $2a$, as shown in Fig. 13.3, should be regarded as a semi-infinite cantilever slab.

The method of analyzing edge-stiffened semi-infinite cantilever slabs is the same as that for the corresponding unstiffened slabs, which was described in Sec. 13.3. However, values of coefficients A' must now be obtained from the charts of Fig. 13.12, and those for B' from the charts of

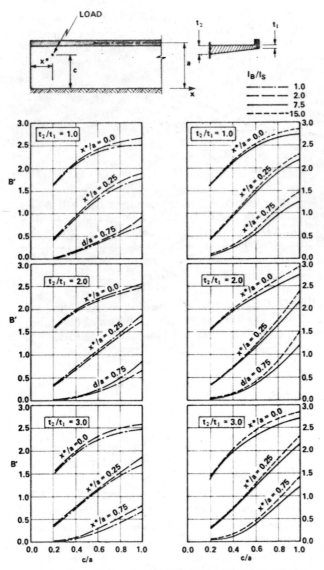

Figure 13.14 Charts for B' for edge-stiffened cantilever slabs.

Fig. 13.14. These charts, which give values of the coefficients for different thickness ratios, different values of I_B/I_S, and various load and reference point locations, are developed in Ref. 1. As was the case in the treatment of unstiffened slabs, the method provides values of M_{yc} only at the root of the cantilever, i.e., at $y = 0$.

In summary, the method consists of the reading of coefficients A' and B' from Figs. 13.12 and 13.14, respectively; the calculation of K from Eq. (13.4); and the determination of M_{yc} from Eq. (13.3).

Example 13.4 Details of this example, chosen to illustrate the use of the method, are shown in Fig. 13.15.

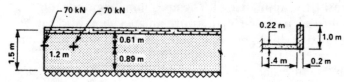

Figure 13.15 Edge-stiffened cantilever slab of semi-infinite length under Ontario wheel loads. (See Fig. 13.7 for notation.)

The load case is the same as that for Example 13.2 and as shown in Fig. 13.9c. The added feature is the stiffening of the longitudinal free edge of the slab. The value of M_{yc} at $x = 0.0$ is sought. With the notation already established,

$$I_B = \frac{0.2 \times 1.0^3}{12} = 16.67 \times 10^{-3} \text{ m}^4$$

$$I_S = \frac{1.4 \times 0.22^3}{12} = 1.24 \times 10^{-3} \text{ m}^4$$

Hence I_B/I_S is 13.4. The value of A' for both loads is found to be 0.83 from Fig. 13.12. From Fig. 13.14, B' for the load on the transverse free edge is found to be 2.5, and for the other load (for which x^*/a is equal to 0.8) is found to be 0.6. It is not necessary to calculate the value of K for either load, since $\bar{x}$ is equal to zero. Hence

$$M_{yc} = \frac{70(0.83)}{\pi}(1 + 2.5)$$

$$+ \frac{70(0.83)}{\pi}\left[\frac{1}{\cosh(0.83 \times -1.2)/1.5} + 0.6\right]$$

$$= 64.7 + 26.1$$

$$= 90.8 \text{ kN} \cdot \text{m/m} \ (20{,}412 \text{ lb} \cdot \text{ft/ft})$$

By comparing this calculation with that for load case (c) of Example 13.2, it can be seen that M_{yc} due to the load on the transverse free edge is reduced from 83.4 to 64.7 kN · m/m (18,748 to 14,545 lb · ft/ft), while the contribution from the farther load is increased from 23.8 to 26.1 kN · m/m (5350 to 5867 lb · ft/ft). Both these changes arise from the enhanced load distributing properties of the stiffened slab as compared with the unstiffened one.

13.6   Method for Edge Beam Moments in Infinite Cantilever Slabs

When a concentrated load is applied to an edge-stiffened cantilever slab, the edge beam is subjected to sagging moments at some positions along its length and to hogging moments at others. The maximum sagging moment M_S occurs in the beam at $x = 0$, that is, at the position on the edge beam directly opposite the applied load. The maximum hogging moment M_H occurs at a distance x_1, from the load. Approximate values of x_1 can be obtained from Fig. 13.16; this figure also provides values of x_1 for semi-infinite slabs (see Sec. 13.7).

In addition to being influenced by the thickness ratio of the slab and the ratio I_B/I_S, the magnitudes of moments in an edge beam are also

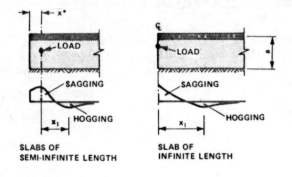

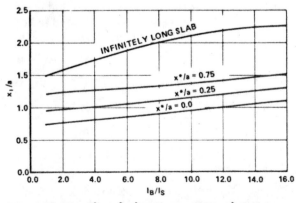

Figure 13.16   Chart for locating maximum hogging moments in edge beams.

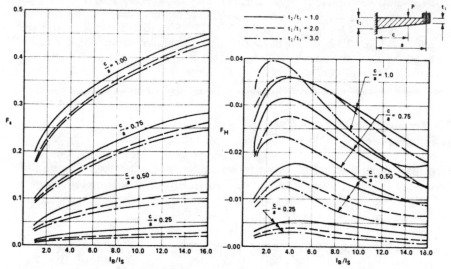

Figure 13.17 Charts of F_s and F_H for cantilever slabs of infinite length.

governed by the width a of the cantilever slab. Moment intensities are directly proportional to a. According to the method given in Ref. 1, values of M_S and M_H due to a single concentrated load can be approximately determined from the following expressions.

$$M_S = PaF_s$$

$$M_H = PaF_H \tag{13.6}$$

where F_s and F_H can be read from the charts given in Fig. 13.17.

The general pattern of moments is as shown in Fig. 13.16. Estimates of edge beam moments for more than one load can be obtained graphically as shown in the following example. It should be emphasized that the methods given in this section and in Sec. 13.7 are intended to provide only reasonable estimates of x_1 with an accuracy of about ± 10 percent. Therefore interpolation based on these approximate values of x_1 are themselves likely to be approximate.

Example 13.5 An edge-stiffened cantilever slab with two loads, as shown in Fig. 13.18a, is analyzed in this example. I_B and I_S are the same as in Example 13.4, so that I_B/I_S is equal to 13.4. From the charts in Fig. 13.17, the values of F_s are 0.27 and 0.14 for c/a equal to 0.75 and 0.50, respectively. Similarly, the corresponding values of F_H are -0.017 and -0.01. By linear interpolation, F_s and F_H are found to be 0.19 and -0.013, respectively, for c/a equal to 0.59. Hence for one 70-kN load

$$M_S = 70(1.5)(0.19) \quad = 20.0 \text{ kN} \cdot \text{m } (14{,}750 \text{ lb} \cdot \text{ft})$$

$$M_H = 70(1.5)(-0.013) = -1.4 \text{ kN} \cdot \text{m } (1030 \text{ lb} \cdot \text{ft})$$

From Fig. 13.16, x_1/a is equal to 2.2. Therefore, x_1 is equal to 3.3 m (10.83 ft). From the above information approximate moment diagrams for the two separate 70-kN loads are plotted in Fig. 13.18b and c. These diagrams are added in Fig. 13.18d to give the approximate maximum value of M_s due to the two loads. As can be seen in the figure, the hogging moments are very small and can be neglected.

13.7   Method for Edge Beam Moments in Semi-infinite Cantilever Slabs

The maximum sagging moment in an edge beam due to a single load occurs at the position $x = 0$ even when the load is in the vicinity of a

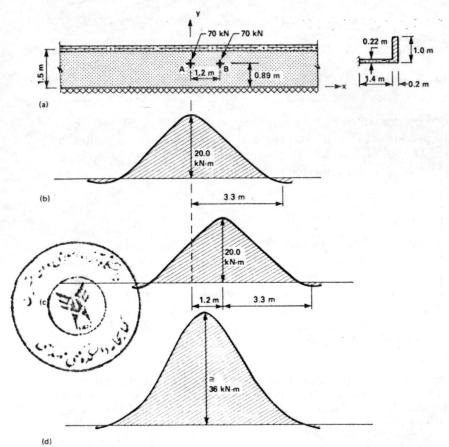

Figure 13.18 Edge beam moments in a cantilever slab of infinite length. (a) Slab details. (b) Edge beam moments due to load A. (c) Edge beam moments due to load B. (d) Edge beam moments due to loads A and B.

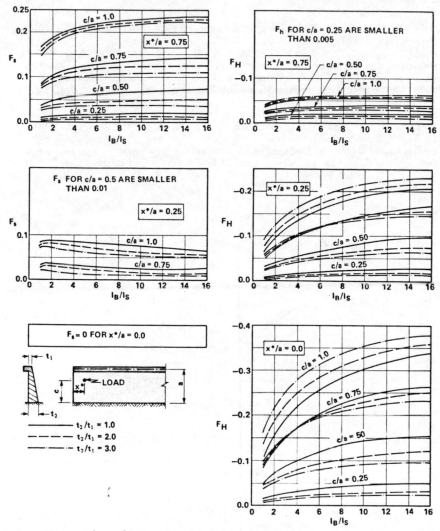

Figure 13.19 Charts for F_s and F_H for cantilever slabs of semi-infinite length.

transverse free edge. The approximate location of the maximum hogging moment can be obtained with the help of Fig. 13.16. The method of determining the maximum values of moments in the edge beam is the same as that for cantilever slabs of infinite length, except that the values of the coefficients F_s and F_H are obtained from the relevant charts given in Fig. 13.19.

References

1. Bakht, B.: Simplified analysis of edge stiffened cantilever slabs, *Journal of Structural Division, ASCE*, 103(ST3), 1981, pp. 535–550.
2. Bakht, B., Aziz, T. S., and Bartusevicius, K. F.: Manual analysis of cantilever slabs of semi-infinite width, *Canadian Journal of Civil Engineering*, 6(2), 1979, pp. 227–231.
3. Bakht, B., and Holland, D. A.: A manual method for the elastic analysis of cantilever slabs of linearly varying thickness, *Canadian Journal of Civil Engineering*, 3(4), 1976, pp. 523–530.
4. Sawko, F., and Mills, J. H.: Design of cantilever slabs for spine beam bridges: Developments in bridge design and construction, *Proceedings of the Cardiff Conference*, Crosby Lockwood, London, 1971.
5. Sawko, F., and Mills, J. H.: Design of cantilever slabs for spine beam bridges: Discussion, *International Conference on Developments in Bridge Design and Construction*, University College, Cardiff, Wales, 1971.

plates, referred to as the torsional parameter and defined by Eq. (1.11)

α_1 = a characterizing parameter defined by Eq. (1.3)

α_2 = a characterizing parameter defined by Eq. (1.4)

β = the articulated plate parameter defined by Eq. (1.20)

γ = a factor related to bridge width and defined by Eqs. (6.2) and (6.3)

Δ = the error in determination of equivalent span length

Δx = the distance between closely spaced transverse sections

δ = a characterizing parameter for shear-weak plates, defined by Eq. (1.34)

θ = a characterizing parameter for rectangular orthotropic plates, referred to as the flexural parameter and defined by Eq. (1.12); or the angle of inclination of the steel tie of a king-post truss to the horizontal, as shown in Fig. 12-9; or the semiangle of dispersion of a concentrated load through the decking, as shown in Fig. 12.6

λ = a characterizing parameter to account for edge stiffening of plates and defined by Eq. (1.41)

λ_m = a modifier to D values for longitudinal moments to account for increase of longitudinal moment intensity due to transverse cell distortion in cellular structures

λ_s = a modifier to D values for longitudinal shear to account for the increase of longitudinal shear intensity due to transverse cell distortion in cellular structures

μ = a factor related to the bridge width, defined by Eq. (4.9)

ν = Poisson's ratio

ν_c = Poisson's ratio of concrete

ρ = the distribution coefficient

σ = the flexural stress

ϕ = a load function

ω = a characterizing parameter for floor systems of truss and similar bridges, as defined by Eq. (12.1) or (12.6)

INDEX

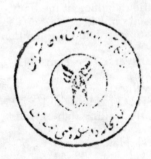